# Decision Analysis
# A Bayesian approach

OTHER STATISTICS TEXTS FROM
CHAPMAN AND HALL

**The Analysis of Time Series**
C. Chatfield

**Statistics for Technology**
C. Chatfield

**Applied Statistics**
D. R. Cox and E. J. Snell

**Introduction to Multivariate Analysis**
C. Chatfield and A. J. Collins

**An Introduction to Statistical Modelling**
A. J. Dobson

**Introduction to Optimization Methods and their Application in Statistics**
B. S. Everitt

**Multivariate Statistics—A Practical Approach**
B. Flury and H. Riedwyl

**Multivariate Analysis of Variance and Repeated Measures**
D. J. Hand and C. C. Taylor

**Multivariate Statistics Methods—a primer**
Bryan F. Manley

**Statistical Methods in Agriculture and Experimental Biology**
R. Mead and R. N. Curnow

**Elements of Simulation**
B. J. T. Morgan

**Essential Statistics**
D. G. Rees

**Applied Statistics: A Handbook of BMDP Analyses**
E. J. Snell

**Elementary Applications of Probability Theory**
H. C. Tuckwell

**Intermediate Statistical Methods**
G. B. Wetherill

*Further information on the complete range of* Chapman and Hall *statistics books is available from the publishers.*

# Decision analysis
# A Bayesian approach

J. Q. Smith
University of Warwick

London New York
CHAPMAN AND HALL

First published in 1988 by Chapman and Hall Ltd.

11 New Fetter Lane, London EC4P 4EE
Published in the USA by Chapman and Hall
29 West 35th Street, New York NY 10017



Printed in Great Britain
by J.W. Arrowsmith Ltd, Bristol

ISBN 0 412 275 10 4 (Hb.)
0 412 275 20 1 (Pb.)



British Library Cataloguing in Publication Data

Smith, J.G.
Decision analysis: a Bayesian approach.
1. Management 2. Bayesian statistical decision theory
I. Title
658.4'033 HD38

ISBN 0-412-27510-4
ISBN 0-412-27520-1 (Pbk.)

Library of Congress Cataloging in Publication Data

Smith, J.Q., 1953-
Decision analysis.

Bibliography: p.
Includes index.
1. Bayesian statistical decision theory. I. Title.
QA279.5.S63 1987 519.5'42 87-15762

ISBN 0-412-27510-4
ISBN 0-412-27520-1 (pbk.)

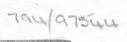

# Contents

**This book is dedicated to the three fathers of modern Bayesian decision analysis:**

**Bruno DeFinetti,**

**Howard Raiffa**

**and**

**Dennis Lindley.**

# Preface

Bayesian decision analysis is a useful tool to many professionals: operations researchers, statisticians, businessmen, economists, engineers, psychologists and computer scientists. This short introductory text is designed for an audience who have attended or are attending a first degree course in a maths-related subject and is based on material given in lectures by me over the last eight years.

For the material in the first five chapters the reader will need as a prerequisite only a first course in discrete probability. The last two chapters are of a more statistical emphasis and assume the reader is acquainted with joint distributions of absolutely continuous random variables.

The text is suitable as a *course book* in decision analysis for penultimate and final-year mathematics undergraduates and final-year undergraduates of statistics, and postgraduate students of economics, operations research, management science, engineering, and computer science. It is suitable as an *introductory text* in Bayesian decision analysis for businessmen who would like to use a good mathematical training to help express and then solve the difficult decision problems that they sometimes face. It can also be used to provide further reading material for undergraduates of economics, mathematical psychology, engineering and computer science attending a course on decision analysis.

In order to emphasize the problem-solving aspects of decision analysis I have tried to keep technical material to a minimum. There are well over 50 exercises and almost 30 worked examples contained in this book, all given to illustrate the various practical and computational techniques of Bayesian problem solving.

Obviously it is extremely important to understand *when* it is appropriate to use a particular problem-solving technique. Sometimes this understanding can only be achieved by knowledge of the precise mathematics underpinning this technique. If this has been the case, I have included the necessary mathematics.

Sections and exercises which are most challenging are labelled by a star. These may be omitted on first reading.

Because of lack of space many topics covered in this book are discussed very briefly. However, I have attempted to give up-to-date references (mostly of review articles and monographs) so that a reader who becomes interested in such a topic can investigate it further.

In the original draft of this book I continually changed the sex of the decision analyst and client. I was informed that this made extremely confusing reading. In this version therefore, the client is always male, and you, the decision analyst, female. I hope that this will satisfy some of the people some of the time!

I am indebted to many colleagues and friends who helped me to compile the material for this book. To name a few: Larry Phillips, Phil Dawid, Tony O'Hagan, Jeff Harrison, Dennis Lindley, Morris DeGroot, Bob Oliver, Dick Barlow, Michael Goldstein, Jim Zidek, Simon French, Kevin McConway and last, but not least, my wife. My thanks to them all.

J. Q. Smith

# 1

# The rudiments of decision analysis

## 1.1 INTRODUCTION

Decision analysis provides the apparatus for resolving business problems logically. Its goal is the identification of a decision which is expected to best satisfy the stated objectives of a client.

A client will employ a decision analyst when he is uncertain about how to act. This uncertainty will usually exist because his problem is not deterministic but probabilistic in nature. He will usually be able to provide you with certain as yet ill-defined objectives and ideas about the consequences of various actions. He should also be able to give you various pieces of information: some *hard* (in the sense of being the result of controlled experiments relevant to the problem at hand, for example the result of a clinical trial or a sample survey) and some *soft* (being the result of his experience and beliefs about how he expects the future to unfold).

Your first task as a decision analyst will be to help your client *organize* all his information into a coherent picture of his problem. You will then need to help him to utilize his information so as to:

1. calculate his *best course of action* given his stated beliefs and objectives;
2. communicate with others *why* he believes his chosen course of action to be optimal;
3. provide a framework within which his ideas can be *critically appraised and modified*, especially in the light of new information not originally incorporated in his model.

In this text we shall concentrate most of our attention on the analysis of decision problems where the client has little or no hard information about some aspect of his problem. In practice you will find that this encompasses most of the problems you are likely to encounter.

An integral part of analysing problems of this kind is the decision analyst's ability to:

(a) identify the *objectives* of an analysis;
(b) identify those *decisions* which are viable;

(c) quantify any *uncertainty* in the problem;
(d) quantify the *costs* of taking any viable decision given each possible eventuality that might arise.

The ways in which these tasks can be accomplished for a given problem together with general methods of achieving objectives 1, 2 and 3 are given in Chapters 1–5. Chapter 6 follows through this approach and introduces useful techniques in Bayesian modelling for the representation and manipulation of more complicated belief structures about uncertain quantities. Chapter 7 shows how to find optimal decisions when a client's belief structure may be complex but his class of viable decisions is analogous to a class of estimates of uncertain quantities.

In this introductory text I have tried to present the rationale behind Bayesian decision-making methodology in a clear and concise way. The techniques I shall develop are illustrated throughout by many simple examples rather than by a few full-blooded case studies. Of course in practice interesting decision analysis rarely solves such simple problems. Perforce, when presented with a choice, I have therefore tended to trade clarity for realism. This is done in the belief that the budding decision analyst will be most able to extrapolate the Bayesian methodology to solve her own (possibly more complicated) problems if these techniques are illustrated simply.

Many aspects of practical decision analysis can only be learned through experience and not study. However, there are many exercises given in this book which if attempted should help the reader to appreciate more readily some of the practical difficulties of applying the Bayesian paradigm to solve problems.

I have included here only material which I believe is of practical importance in instructing the decision analyst to promote wise action on behalf of her client. On the other hand, where necessary, I have given full logical/mathematical arguments for how I believe a decision analyst should proceed in solving her client's problem.

One of the simplest types of solution to decision problems will be illustrated in the remainder of this chapter. Some simple 'textbook' decision problems will be solved without the use of some of the more sophisticated techniques of decision analysis. As the book progresses I shall show how to solve more technically difficult problems and discuss how best to use the solutions of textbook problems to give 'sensible' (if not optimal) solutions to complicated practical problems.

Let us begin therefore by setting up some notation and giving some definitions.

## 1.2 NOTATION AND DEFINITIONS

Let $\Theta$ denote the space of all possible outcomes $\theta$ of the uncertainty in your problem and $D$ denote the space of all viable decisions $d$ to the problem.

Throughout the first five chapters we shall assume that $\theta$ is discrete (i.e. contains at most a countable number of points). This assumption, although not crucial to the development given in later chapters, makes the explanation of the rationale behind decision analysis much easier to follow.

As stated above, to analyse a decision problem we must first quantify the objectives of the analysis and the client's knowledge about the problem. To be precise, we need:

1. A *loss function* $L(d, \theta)$ specifying (usually in monetary terms) how much a client will lose if he made a decision $d \in D$ and the outcome were $\theta$. The function $L(d, \theta)$ needs to be known for all values $d \in D$ and $\theta$.
2. For each decision $d \in D$ a *probability mass function* $p(\theta | d)$ on $\theta \in \Theta$ with

$$p(\theta | d) \geqslant 0 \text{ for all } \theta \in \Theta \quad \text{and} \quad \sum_{\theta \in \Theta} p(\theta | d) = 1$$

The function $p(\theta | d)$ gives the probability that $\theta$ will take its various values given that a client chooses a decision $d \in D$. If you have used data to assess this mass function it is often called your *posterior mass function*; otherwise it is called a *prior mass function*.

**Notes** A client's losses might be negative and hence they would be gains. For example, $L(d, \theta) = -1$ would imply that he would gain \$1 if he took decision $d$ and the outcome was $\theta$. Statisticians, being pessimistic people, usually work with losses. On the other hand, economists, perforce optimistic, prefer to work with pay-off's. In this book I will be pragmatic. If a problem deals mostly in gain we will use a *pay-off function* $R(d, \theta) = -L(d, \theta)$ instead of a loss function. Otherwise I will stick with the loss function.

Suppose you have managed to quantify your client's mass function and loss function. What algorithm should you employ to find his 'optimal decision'? For many problems he would like to choose a decision which will minimize the loss he expects to result from his action. The client's objective can then be stated in terms of the following algorithm.

*The expected monetary value (EMV) algorithm*

A decision maker should choose a decision $d^* \in D$ which minimizes his expected loss (or, equivalently, maximizes his expected pay-off), the expectation being taken across the outcome space $\Theta$.

By the definition of mathematical expectation, to follow such an algorithm the decision-maker minimizes over $d \in D$

$$L(d) = \sum_{\theta \in \Theta} L(d, \theta) p(\theta | d)$$

where $L(d)$ denotes his expected loss when he takes decision $d \in D$; or equivalently, maximizes over $d \in D$

$$R(d) = \sum_{\theta \in \Theta} R(d, \theta) p(\theta | d)$$

where $R(d)$ denotes his expected pay-off when he takes decision $d \in D$.

A decision $d^*$ which minimizes $L(d)$ or maximizes $R(d)$ over $d \in D$ is commonly called a *Bayes decision.* A Bayes decision will clearly be an optimal decision if the EMV algorithm relates to the client's true objective.

## 1.3 EXAMPLES

Here are three idealized problems which are solved using the EMV algorithm.

*Example 1.1*

A manager must decide whether to lease a small machine at a total cost of \$10 000 (decision $d = 1$) or a large machine at a cost of \$30 000 (decision $d = 2$). The small machine can produce at most 500 products a week and the large machine at most 2000. The lease for each machine will last for 50 weeks.

He hopes to find an outlet for his product through a large wholesaler. If this happens (outcome $\theta = 2$) he would expect to be able to sell 1800 items per week. If he could not obtain this outlet, however (outcome $\theta = 1$), he would only be able to sell 400 items per week. If he makes \$1 profit on each item he sells, find his Bayes decision.

**Answer** Let $p$ denote his probability that $\theta = 2$. His pay-off $R(d, \theta)$ associated with decision $d$ and outcome $\theta$ is the number of items he sells over the 50 weeks' operation minus the cost of any machine he leases.

Thus, calculated in \$,

$$\begin{aligned} R(1,1) &= 400 \times 50 - 10\,000 = 10\,000 \\ R(1,2) &= 500 \times 50 - 10\,000 = 15\,000 \\ R(2,1) &= 400 \times 50 - 30\,000 = -10\,000 \\ R(2,2) &= 1800 \times 50 - 30\,000 = 60\,000 \end{aligned}$$

It follows that his expected pay-offs $\bar{R}(1)$ and $\bar{R}(2)$ associated with decisions $d = 1$ and $d = 2$ are

$$\begin{aligned} \bar{R}(1) &= 10\,000(1-p) + 15\,000p = 5000(2+p) \\ \bar{R}(2) &= -10\,000(1-p) + 60\,000p = 10\,000(7p-1) \end{aligned}$$

Clearly $\bar{R}(1) > \bar{R}(2)$ if and only if

$$\begin{aligned} & 5000(2+p) > 10\,000(7p-1) \\ & \Leftrightarrow p < 4/13. \end{aligned}$$

$\bar{R}(1)$ and $\bar{R}(2)$ are plotted in Fig. 1.1.

By following the EMV algorithm we can now see that the manager's Bayes decision rule is:

Lease the small machine if he believes that the probability of obtaining his

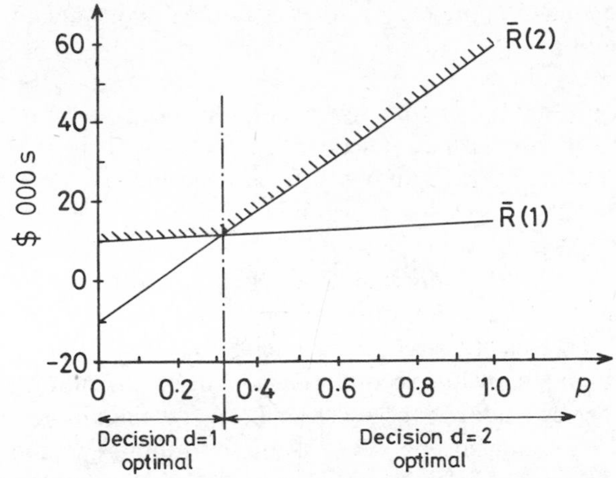


**Figure 1.1** Expected pay-offs $\bar{R}$ associated with the purchasing decisions $d = 1, 2$ in Example 1.1

outlet is less than 4/13 and lease the large machine otherwise. (If $p = 4/13$ then each decision is equally good.)

Notice in this example that a decision-maker is required to assess the probability of an uncertain event even when he has no data to help him. The EMV algorithm often needs *subjective probability statements* from a client before it can identify an optimal decision. Note, however, that such a statement does not need to be precise. In this problem all the client need state is whether his subjective probability is less or greater than 4/13. You can then identify the best course of action for him.

*Example 1.2*

A medical laboratory has to test $N$ samples of blood to see which have traces of a rare disease. The probability any one patient has the disease is $p$, and given $p$, the probability of any group of patients having the disease is uninfluenced by the existence or otherwise of the disease in any other disjoint group of patients. Because $p$ is believed to be small it is suggested that the laboratory combine the blood of $d$ patients into equal sized pools of $n = N/d$ samples where $d$ is a divisor of $N$. Each pool of samples would then be tested to see if it exhibited a trace of the infection. If no trace were found then the individual samples comprising the group would be known to be uninfected. If on the other hand a trace were found in the pooled sample it is then proposed that each of the $n$ samples comprising that pool be tested individually.

If it costs \$1 to test any sample of blood, whether pooled or unpooled, find the Bayes decision for the optimal size $d$ of the groups of patients for a given value of $p$.

**Answer** You are given that the space of decisions you are to consider is the set of divisors of $N$. All the uncertainty in the experiment exists because you do not know $\theta(d)$ – the number of tests your client will need to do if he chooses to pool the samples into groups of $d$ samples. His monetary loss is just $L(d) = \$\theta(d)$.

The uncertain quantity $\theta(d)$ can be broken down into two components, being the sum of the number $n$ of tested pools plus the number of individual patients that subsequently need to be checked. If $d = 1$, and he chooses to test samples individually, the second component of this sum is known to be zero. So the corresponding expected loss $L(1) = N$, the number of patients.

Suppose $d > 1$ and you choose to pool the samples in some way. If $\lambda$ denotes the probability that a pool has no trace of diseased blood, then $\lambda$ is the probability that no patient in the pool has the infection. Since patients have the disease independently it follows by the laws of probability that

$$\lambda = (1 - p)^d$$

So, since he will test $n = N/d$ such pooled samples, the expected number of pooled samples that need retesting is $n(1 - \lambda)$. If a pool is found to have traces of the disease then all members of the pool will be retested. So the expected number of samples that subsequently need retesting is

$$dn(1 - \lambda) = N\{1 - (1 - p)^d\}$$

Adding this number to the chosen number of pools ($n = N/d$) gives the expected number of tests (or equivalently the expected loss in \$) for choosing to use sample pools of size $d > 1$. Combining these results gives that

$$\bar{L}(d) = \begin{cases} N & d = 1 \\ N\{d^{-1} + 1 - (1 - p)^d\} & d > 1,\ d \text{ a divisor of } N. \end{cases}$$

Although $\bar{L}(d)$ is not a linear function of $d$ (as it was in our first example), given $p$, $\bar{L}(d)$ can be easily calculated for each divisor $d$ of $N$ and the decision which minimizes $\bar{L}(d)$ found. In Exercise 1.2 at the end of this chapter you are asked to show that if $p \geqslant 0.31$ you should choose to test samples individually. On the other hand, if $N$ is divisible by 3 and $p < 0.31$ then it is always optimal to pool samples in some way.

In the above example your client's uncertainty was affected by the decision he chose. Even in simple problems, whenever the probabilities of uncertain events depend on components of decisions you need to make, it may be difficult to see how to break down the problem in terms of decision spaces, random variables and losses so that the EMV algorithm can be followed. Therefore it is very useful to have a pictorial representation of the components of a problem. One commonly used representation, the decision tree, is discussed in the next chapter. More sophisticated diagrams of problems are given in Chapter 4 (causal diagrams) and Chapter 5 (influence diagrams).

Notice that now you have performed this analysis the decision space $D$ you initially chose seems rather artificial. The optimal policy seems to depend rather too heavily on the divisors of $N$. Intuitively if $N$ is large you might expect to act similarly if you have one extra patient (say). This can be overcome by allowing pools of only approximately the same number of patients.

Furthermore, if $p$ is small surely there is a case for allowing sub-pools of samples to be taken when a large pool indicates the presence of a disease.

The purpose of decision analysis is not only to find the best alternative from a given set of decisions $D$ but also to provide the decision-maker with a framework from which he might perceive previously unconsidered alternative policies. To assess clearly the efficacy of complicated options you usually first need a precise appreciation of the merits and demerits of simpler options.

The next example gives another simple non-linear decision problem solved by the EMV algorithm.

*Example 1.3*

Items of a certain type that your client manufactures are independently either imperfect (with probability $\pi$) or perfect. If he dispatches an imperfect item he will lose a valuable client at an expected cost to him of \$10 000. He can choose between:

decision $d = 1$ – to dispatch an item without checking it
decision $d = 2$ – before dispatching an item he inspects it with a foolproof quality control. If that item is found to be faulty he will replace it by another item which he will again check. He continues to replace items in this way until he finds one with no imperfection.

The cost of making any one item is \$3000 and each inspection costs \$1000. Find your client's Bayes decision as a function of $\pi$.

**Answer** The expected loss $\bar{L}(1)$ associated with $d = 1$ is the cost of making the item plus the expected cost of losing your client's customer. Hence (in \$)

$$\bar{L}(1) = 3000 + 10\,000\pi$$

On the other hand the expected loss $\bar{L}(2)$ associated with $d = 2$ is the cost of making and checking the $r$ items (say) before finding the first perfect item. How many items $R$ you need to check before a perfect one is found is uncertain but clearly, by the stated independence,

$$P(R = r) = \pi^{r-1}(1 - \pi)$$

Since by the above argument

$$L(2|R=r) = r(1000+3000) = 4000r,$$

$$\bar{L}(2) = 4000 \sum_{r=1}^{\infty} r\pi^{r-1}(1-\pi)$$

$$= 4000(1-\pi) \sum_{r=1}^{\infty} r\pi^{r-1}$$

$$= 4000(1-\pi)(1-\pi)^{-2}$$

$$= 4000(1-\pi)^{-1}.$$

So you should prefer $d=2$ to $d=1$ when $\bar{L}(2) < \bar{L}(1)$,

i.e. $4000(1-\pi)^{-1} < 3000 + 10\,000\pi$

i.e. $10\pi^2 - 7\pi + 1 < 0$

i.e. $(5\pi - 1)(2\pi - 1) < 0 \qquad 0 < \pi < 1$

i.e. when $0.2 < \pi < 0.5$

As in the first example your optimal decision will depend upon how the client chooses to specify the probability. Notice that although $d = 1$ is optimal when $\pi$ is close to one, the associated loss is extremely high. In the light of this he might, in practice, consider taking a third type of decision – for example, spending money improving the manufacturing process. Thus we again see that by quantifying your client's problem you can encourage him to consider other options that before your analysis might have appeared unnecessary.

## EXERCISES

1.1 (on Example 1.1) Assume that the manager has an additional decision ($d = 3$) where he can resort to industrial espionage at a cost of \$10 000 to discover whether or not the wholesaler will provide an outlet for his goods. Find his new Bayes decision rule.

1.2 (on Example 1.2) Show that $d^{-d^{-1}}$ is increasing when $d > e$. Show that you should never pool unless

$$1 - p \leqslant \min\{(\tfrac{1}{2})^{1/2}, (\tfrac{1}{3})^{1/3}, (\tfrac{1}{4})^{1/4}\}.$$

1.3 (on Example 1.2) If the total number of samples $N$ is divisible by 4 show that, under the EMV strategy, it is always better to pool into groups of 4 rather than 2 whatever is the value of $p$.

1.4 (on Example 1.2) If $N$ is prime find your Bayes rule explicitly.

1.5 (on Example 1.3) Suppose the loss of the customer costs you \$$A$, an inspection costs you \$$B$ and the cost of making item is \$$C$ (rather than the amounts given in original example). Show:

(i) if $A/B < 4$ you never inspect regardless of the value of $\pi$;
(ii) if $(A^2 + B^2) > A(C + 2B)$ you should inspect for some value of $\pi$;
(iii) if $B = 0$ you should inspect if and only if $\pi < 1 - C/A$;
(iv) if $C = 0$ you should inspect if and only if

$$(\pi - \tfrac{1}{2})^2 < \tfrac{1}{4} - B/A.$$

(This illustrates that you need not require explicit values of losses before you can identify an optimal policy.)

1.6 (on Example 1.3) Re-analyse Example 1.3 on the assumption that your inspection is not totally precise, but says 'imperfect' when an item is imperfect with probability 0.9 and says 'perfect' when an item is perfect with probability 0.9.

1.7 The effect of $d$ kilos of fertilizer on the expected yield $\lambda$ of a crop is given by the equation

$$\lambda = 10(8 + \sqrt{d}) \qquad d < 1000$$

The cost of a kilo of fertilizer is \$5 and the profit on a unit of crop is \$10. Find the Bayes decision for $d$, given $d < 1000$, together with the expected pay-off under this optimal decision.

1.8 You believe that a coin is fair with probability 0.7 and double-headed with probability 0.3. You can guess immediately whether or not the coin is fair or buy the results of $n$ tosses of the coin $n = 1, 2, 3, \ldots$ before you make your decision. Each toss will cost you \$1. You lose \$100 if you say the coin is fair and it is not. You gain \$50 if you make the right guess, and lose nothing if you say that the coin is unfair when it is fair. Find your Bayes decision and its associated expected pay-off.

1.9* A batch of items has two imperfect items in it with probability $p$ and no imperfect items with probability $1 - p$. If the batch is sent off to a customer with any imperfect item in it the company will lose its contract (at an expected cost of $C_1$). To clean the machine costs $C_2$ and this will ensure that the next batch of $N$ items is perfect. The company can either:
(a) Clean the machine immediately (decision $d = N + 1$).
(b) First take a random sample of $r$ items at a cost of $rc$ from the batch (decision $d = r$). If these test items are found to be perfect the batch will be dispatched, otherwise the machine will be cleaned. Advise the company on its best course of action.

# 2

# Decision trees

## 2.1 INTRODUCTION

In the introduction of Chapter 1, I stated that the first task of a decision analyst was to try to obtain a coherent picture of her client's problem. In very simple problems, like Examples 1.1, 1.2 and 1.3, this may well be a trivial exercise but once problems start becoming more complex it can be very difficult to specify and quantify the relationships between decisions and uncertain events.

One well-known pictorial representation of a client's problem is the decision tree. Once drawn, the decision tree can be used to help calculate the expected pay-off associated with each sequence of decisions open to the client. It is then possible to identify the Bayes decision to a given problem.

The technique of drawing and 'folding back' a decision tree is most easily understood through an example. The following well-posed, if artificial, example is used to illustrate the method, and most of this chapter is devoted to its solution.

*Example 2.1*

The government has given an oil company the option to drill in either field $A$ or field $B$ but not both. The probabilities that oil is present in $A$ and in $B$ are 0.4 and 0.2 respectively and these two events are independent. A net profit of \$77 million is expected if oil is struck in $A$ and a net profit of \$195 million is expected if oil is struck in $B$.

The decisions open to the company are:

(i) not to accept the option to drill in either area;
(ii) to accept the option and drill in either area $A$ or area $B$ immediately;
(iii) to pay for the investigation of one of the fields (but not both) and in the light of the information thus obtained, to choose between decisions (i) and (ii) above. The investigation will either advise drilling or not.

The investigative procedure, although costly, is not totally reliable. If oil is present in a field then the investigators will advise drilling with probability 0.8. If oil is not present, then the investigators will advise drilling with probability 0.4. The cost of accepting the option and drilling the chosen field is \$31 million while the cost of investigating a field is \$6 million.

Draw a decision tree and advise the company on the best decision rule under the expected monetary value algorithm. What is the expected pay-off to the company if they use this decision rule?

When given information in this form your first step should be to systematically tabulate all the information you have. In this example, let $A$ and $B$ denote the presence of oil in fields $A$ and $B$ respectively. Let $\bar{A}$ and $\bar{B}$ denote the complementary event to $A$ and $B$. The events labelled $a$ and $b$ occur when the investigation advises drilling in respective fields $A$ and $B$. The events labelled $\bar{a}$ and $\bar{b}$ will denote their complements.

**Probabilities given in the problem** Using the notation given above you have been given that

$$P(A) = 0.4, \quad P(B) = 0.2$$
$$P(a|A) = 0.8, \quad P(b|B) = 0.8, \quad P(a|\bar{A}) = 0.4, \quad P(b|\bar{B}) = 0.4$$

You are also told that $A$ and $B$ are independent, so clearly the result of an investigation of one field gives no information about the other field. So, for example, $P(a|A, B) = P(a|A)$. Formally stated, conditional on $A$ or $\bar{A}$, $a$ is independent of $B$, and conditional on $B$ or $\bar{B}$, $b$ is independent of $A$.

Thus we have, for example,

$$P(a \cap A \cap B) = P(a|A \cap B)P(A \cap B)$$
$$= P(a|A)P(A \cap B)$$

by the conditional independence given above

$$= P(a|A)P(A)P(B)$$

by the independence of $A$ and $B$. All probabilities relevant to your problem can be discovered using analogous formulae and are tabulated in Table 2.1.

**Table 2.1** The joint probabilities of all events of interest in Example 2.1

| Joint probabilities | $A \cap B$ | $\bar{A} \cap B$ | $A \cap \bar{B}$ | $\bar{A} \cap \bar{B}$ | Marginal probabilities |
|---|---|---|---|---|---|
| $a$ | 0.064 | 0.048 | 0.256 | 0.192 | 0.56 |
| $\bar{a}$ | 0.016 | 0.072 | 0.064 | 0.288 | 0.44 |
| $b$ | 0.064 | 0.096 | 0.128 | 0.192 | 0.48 |
| $\bar{b}$ | 0.016 | 0.024 | 0.192 | 0.288 | 0.52 |
| Marginal probabilities | 0.08 | 0.12 | 0.32 | 0.48 | |

**Table 2.2** Pay-offs (in $ m) – profit less drilling cost

| | Action | | |
|---|---|---|---|
| Event | Drill $A$ | Drill $B$ | Don't drill |
| $A \cap B$ | 46 | 164 | 0 |
| $\bar{A} \cap B$ | −31 | 164 | 0 |
| $A \cap \bar{B}$ | 46 | −31 | 0 |
| $\bar{A} \cap \bar{B}$ | −31 | −31 | 0 |

**Pay-offs given in this problem** In this problem there are three *terminal* decisions – that is, decisions which can be acted upon after choosing any preliminary investigation. These will be to drill $A$, to drill $B$ or not to drill either field. The pay-off when making each decision given the states of the two fields is the profit less any drilling cost, and are summarized from the question in Table 2.2.

Finally, the cost of any investigation is $6m. Having summarized your information you are now ready to draw your tree.

## 2.2 HOW TO DRAW A DECISION TREE

(i) Find a large piece of paper (the back of computer print-out is very useful for drawing decision trees). Work from the left to the right of the page. First identify the set of decisions $D_1$ your client needs to make before he can observe any of the outcomes of interest. In this example these decisions are:

$d_1$ – to investigate field $A$
$d_2$ – to investigate field $B$
$d_3$ – to drill $A$ without investigation
$d_4$ – to drill $B$ without investigation
$d_5$ – to not investigate and to drill neither field.

Whenever your client has a choice between a set of decisions such as $D_1$, draw a square box to represent $D_1$ and a fork emerging from that box for each possible decision. Each such square box is called a *decision node* and each such fork is called a *decision fork*.

Thus in this example, start drawing your decision tree as in Fig. 2.1.

(ii) Given that your client takes any one of these initial decisions he may then observe some event which was initially uncertain to him. In this example, after taking decision $d_1$ to investigate field $A$ he will observe either a positive ($a$) or negative ($\bar{a}$) recommendation to drill. If some such uncertainty is resolved after taking a decision, draw a circle (called a *chance* node) at the

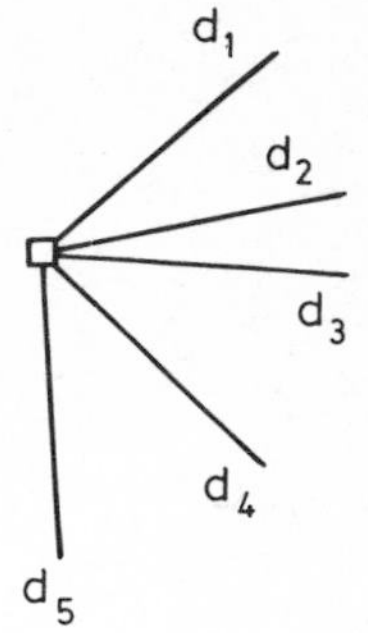


**Figure 2.1**

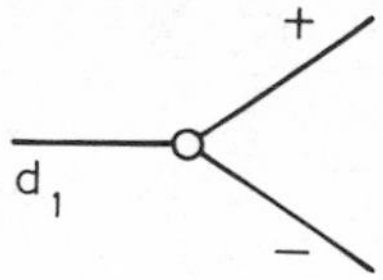


**Figure 2.2** One of our initial chance nodes

right-hand end of your decision fork with forks (called *chance forks*) emerging to the right of it, each fork labelled by one of the possible outcomes. The first chance node to appear after decision $d_1$ of the example is drawn in Fig. 2.2.

(iii) You now draw on to the tree any further decision node that can represent a decision which needs to be made contingent only on decisions and outcomes that are already represented in the tree so far. For example, in the case above, given $\bar{a}$ your client must choose between drilling $A$, drilling $B$ or not drilling. Draw another decision node with three forks out of it, labelled by these options. In general, include all logically feasible actions on the tree whether or not they, at first glance, seem sensible.

Similarly, you can add further chance nodes to the tree when the corresponding events have probabilities which only depend on decisions and outcomes you have already depicted and whose uncertainty is resolved after these previous decision and outcomes have occurred.

By adding nodes to the tree in this way you will eventually reach a point where no more nodes and decisions can be added using these rules. There may, however, be some residual uncertainty in the problem. In this example, after ($d_1$, $a$, drill $A$), there still remains to represent whether oil exists in $A$ or not, and this will effect your expected pay-off. In this case this uncertainty can be represented by a single outcome node with two branches denoted by $A|a$ and $\bar{A}|a$. Adding sufficient outcome nodes to represent all your client's residual uncertainties, you will obtain the completed skeleton of your tree representing Example 2.1 as given in Fig. 2.3.

Now you have a diagram of your client's problem you should pause to ask yourself the question: 'Have I included all his possible options and included all aspects of uncertainty pertinent to choosing between those options?' Having drawn your tree you will often find that this is not the case. If you believe you have missed out some promising looking decisions or some large component of variation, add them to your model now by adding

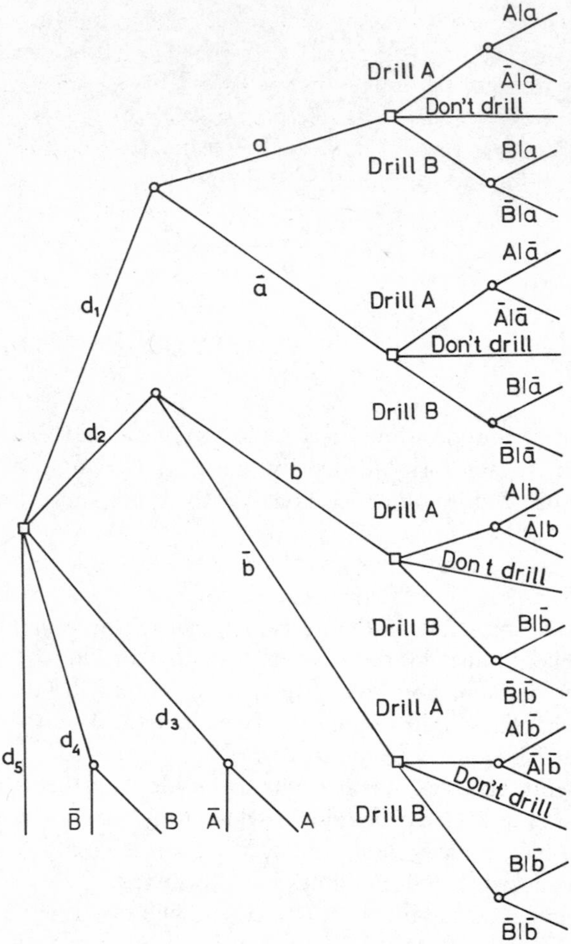


**Figure 2.3** A tree skeleton

more decision nodes and forks and chance nodes and forks to your tree. Do this until you are happy that the tree is a good reflection of the problem in hand. Add to your summary of information all the necessary probabilities and pay-offs associated with these newly included options or components of variation.

(iv) Now your tree is drawn you need to fill your tree with the information provided. Firstly the pay-offs associated with each combination of outcome and decision need to be added to the right-hand tip of the *terminal* forks of your tree – that is, those forks with no node on their right-hand end. Care must be taken to ensure that you include all profits and costs incurred from taking decisions and observing outcomes that lead to that terminal fork. In our example, the pay-offs associated with the terminal nodes are just the corresponding relevant pay-offs obtained from Table 2.2 minus any cost of investigating an oilfield that might be incurred on the way to that terminal fork. Thus the pay-off associated with the sequence deciding to investigate $A$ ($d_1$), getting a positive response ($a$), drilling $A$ and finding oil ($A$) is \$46 m minus the \$6 m cost of investigation, i.e. \$40 m. Write the number 40 on the right-hand tip of the $A|a$ fork. If you proceed in this way for all the terminal forks of Fig. 2.3 you should obtain the 'terminal pay-offs' given in Fig. 2.4.

(v) The last components you need to add to your tree are the probabilities

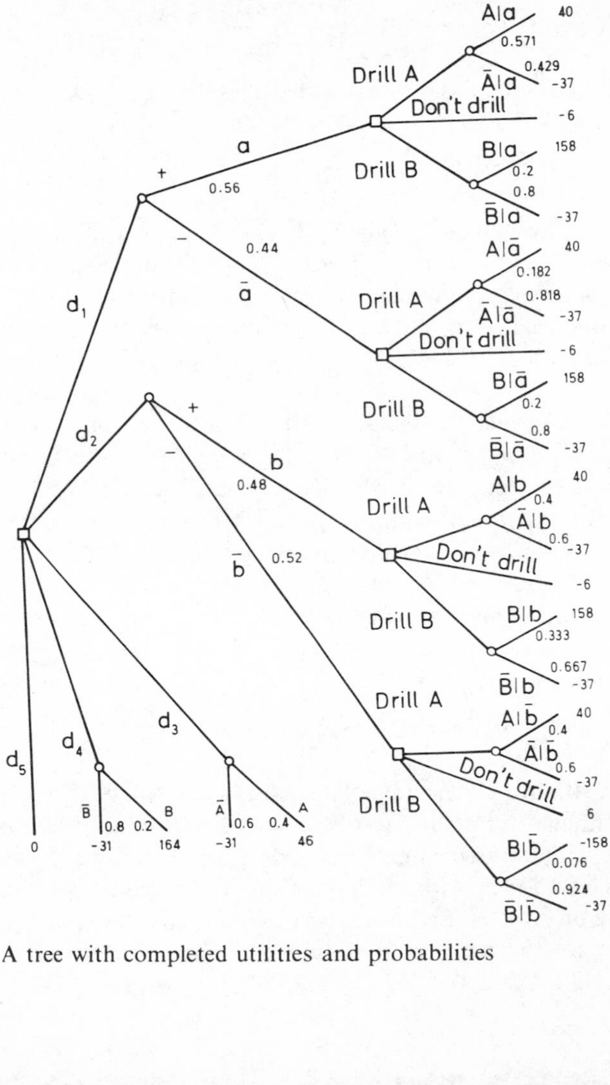


**Figure 2.4** A tree with completed utilities and probabilities

relevant to the chance forks of the tree. In this example, suppose your client is considering decision $d_1$ of whether to investigate oilfield $A$. At the time of this decision your client is uncertain about whether he will observe a positive reading. ($a$) or not ($\bar{a}$). However, you have probabilistic information about the event $a$. From Table 2.1, $P(a) = 0.56$ and $P(\bar{a}) = 0.44$. Write these probabilities over the relevant chance forks. Now let's suppose you observe a positive result and your client then decides to drill field $A$. This is depicted in Fig. 2.3 by the two topmost terminal chance forks. You need to know the probability of $A$ (and hence $\bar{A}$) *given* these previous outcomes and decisions. In this case the only thing that influences this probability is $a$. $P(A|a)$ is not given directly but can be easily calculated, either from the original data using Bayes rule,

$$P(A|a) = \frac{P(a|A)P(A)}{P(a)}$$

or by using the probabilities given in Table 2.1 using the formula

$$P(A|a) = \frac{P(A \cap B \cap a) + P(A \cap \bar{B} \cap a)}{P(a)}$$

In either case you will calculate $P(A|a) = 0.571$ and hence $P(\bar{A}|a) = 1 - P(A|a) = 0.429$. These probabilities can now be added to the uppermost and next to uppermost terminal chance forks respectively.

Probabilities need to be added to each chance fork in this way. If you do this you should obtain the numbers given in Fig. 2.4.

(vi) Your tree now depicts the structure of the problem and contains all information relevant to finding a Bayes decision. Here is the way you identify your client's Bayes decision. It is called 'folding back the tree'.

Work from right to left. If there are any terminal chance forks then you write an expected pay-off over the corresponding chance node to the left of these forks. This expected pay-off is simply the product of terminal pay-off and associated probability summed over this set of forks. For example, over the chance node arising from $d_1$, $a$, Drill $A$, write the number 7.00 where 7.00 is calculated from

$$0.571 \times 40 + 0.429 \times (-37) = 7.00$$

(vii) By doing this for all terminal chance forks you will have a set of decision forks with either a terminal pay-off at its right-hand end (as for $d_5$) or a chance node with an associated expected pay-off. Choose the decision fork from a decision node $D'$ with the largest pay-off/expected pay-off, write this expected pay-off over $D'$ and delete the other forks that emerge out of $D'$. For example, for the decision node after $d_1, a$, the pay-offs are 7 (Drill $A$),

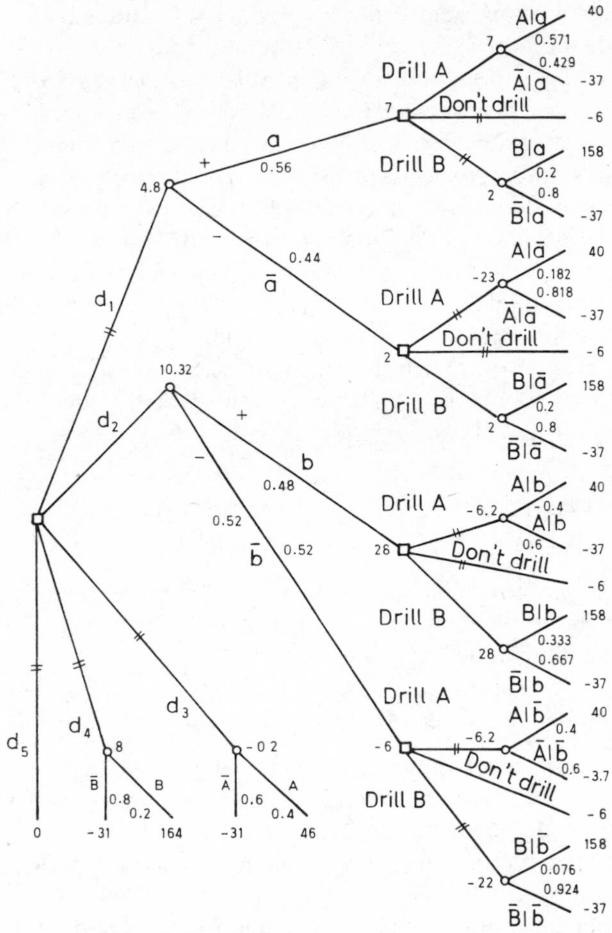


**Figure 2.5** A 'folded back' decision tree

– 6 (Don't drill) and 2 (Drill *B*); so delete the Don't drill and Drill *B* forks and write 7 over the decision node.

(viii) Repeat procedures (vi) and (vii) so that numbers are finally written over nodes connected by a fork to the original decision node. The tree has now been completed. Figure 2.5 gives the decision tree from Example 2.1.

(ix) Your client's optimal decision can now be read from the tree. Pick the decision in the initial decision node on the left with associated fork having the largest expected pay-off. In this example this is decision $d_2$. Provided your client follows the ensuing course of action, his expected pay-off associated with this decision $d_2$ is exactly the expected pay-off written over its right-hand node (\$10.32 m).

Of course it is not enough in this problem to just state that your client should investigate field $B$ (decision $d_2$). Your client also needs to know how to act after making this decision. This can also be read from the tree. In this example, having taken decision $d_2$ you will discover the result of your investigation ($b$ or $\bar{b}$). Given outcome $b$, your client should act on the decision labelled by the undeleted decision fork. In our tree this is the 'Drill $B$' fork. On the other hand if the investigation advises against drilling (outcome $\bar{b}$) your client should take the course of action 'Don't drill either field' labelling the undeleted fork arising from this outcome.

This optimal course of action, which specifies how your client should act contingent on the outcomes of various events which at the time of analysis are uncertain, is called a *Bayes decision rule.*

## 2.3 WHY THIS ALGORITHM FINDS THE DECISION RULE WHICH MAXIMIZES EXPECTED PAY-OFF

Here we interpret what the various forks and nodes represent in a technical sense and hence explain why the algorithm given in the last section actually identifies the optimal decision rule.

For any node $n$ connected on the left to forks $f_1, \ldots, f_m$, the number above the right-hand end of $f_i$ denotes the expected pay-off $G(f_i | \boldsymbol{\theta}, \mathbf{d})$ that is:

(a) conditional on all outcomes $\boldsymbol{\theta}$ occurring and decisions $\mathbf{d}$ made and represented by forks 'preceding' $f_i$, i.e. lying along a line connecting the far left-hand node at the base of the tree to $f_i$;

(b1) conditional on, in the case of $n$ being a decision node, the decision labelled by $f_i$, $1 \leqslant i \leqslant m$;

(b2) conditional on, in the case of $n$ being a chance node, the outcome $\theta_i$ occurring that is labelled by $f_i$, $1 \leqslant i \leqslant m$; and

(c) provided that your client subsequently takes the decisions available to him which will maximize his expected pay-off conditional on $(\boldsymbol{\theta}, \mathbf{d})$.

At a decision node therefore, our algorithm simply picks out the fork denoting the decision which promises the client his maximum expected pay-off provided that he subsequently acts wisely. This expected pay-off is placed over $n$. On the other hand, when $n$ is a chance node our algorithm just calculates a conditional expected pay-off given subsequent wise actions using a well-known property of expectation. Thus the expected pay-off $\bar{R}(\boldsymbol{\theta}, \mathbf{d})$ given $\boldsymbol{\theta}, \mathbf{d}$ (the number to be placed over $n$) can be calculated by the formula

$$\bar{R}(\boldsymbol{\theta}, \mathbf{d}) = \sum_{i=1}^{m} \bar{R}(f_i | \boldsymbol{\theta}, \mathbf{d}) P(\theta_i | \boldsymbol{\theta}, \mathbf{d})$$

where $P(\theta_i|\boldsymbol{\theta}, \mathbf{d})$ is the probability of $\theta_i$ given the outcomes and decision leading to it.

Using the interpretation of the number calculated above, it is not difficult to prove inductively that our algorithm chooses at each decision fork, the decision which maximizes expected pay-off conditional on all preceding outcomes and decisions.

For a more detailed and formal proof that this algorithm maximizes expected pay-off, see Raiffa and Schlaifer (1961).

## 2.4 THE EXPECTED VALUE OF PERFECT INFORMATION

Before drawing a tree it is often a good idea to calculate the expected value of perfect information (EVPI). This is defined as your client's expected pay-off given he knows the state of nature minus his expected pay-off given he performs no experimentation. In our example you can see from Tables 2.1 and 2.2 that his expected pay-off given perfect information about the existence of oil, or otherwise, in the two fields, given that your client chooses a decision to maximize his expected pay-off, is:

$$0.08 \times 164 + 0.12 \times 164 + 0.32 \times 46 + 0.48 \times 0 = 47.5 \qquad \text{(in \$m)}$$

His best expected pay-off given that he does not experiment, i.e. investigate one of the oil fields, is \$8 m, the expected pay-off of drilling *B* immediately. Thus

$$\text{EVPI} = 47.5 - 8 = 39.5 \qquad \text{(in \$m)}$$

Any decision rule which performs an investigation that costs \$$C$ where $C > 39.5$ can be immediately discarded. This is because, even if the investigation provided perfect information about the two fields, the pay-off would still be less than \$8 m and so less preferable to drilling *B* immediately.

If you calculate the EVPI before drawing a decision tree you will often save time analysing decision rules which are clearly not going to be optimal (see Exercise 2.3). The EVPI gives you an upper bound to the amount your client should be prepared to consider paying for information.

## 2.5 PRACTICAL PROBLEMS ASSOCIATED WITH DECISION TREES AND THE EMV ALGORITHM

1. In practical problems it is very difficult to isolate *all* the viable courses of action open to your client. In fact if you were pedantic and tried to analyse every conceivable possible course of action you would never have the time to finish any analysis. In practice you are advised to analyse the decision rules which seem to you likely to be close to optimal schemes. Of course,

there is always the danger if you do this that you will not find the decision rule which would be optimal had you analysed a large class of decision rules. However, this will not matter too much if the expected pay-off from your chosen decision is not much less than the expected pay-off from the optimal decision associated with the larger problem.

2. In practice both the pay-offs associated with various courses of action and outcomes and the probabilities of outcomes are often very difficult to assess. A considerable amount of recent research has been directed into trying to solve these problems and will be discussed in much more detail in Chapter 4 and 5. Of course, if the values you assign to your client's pay-offs and probabilities are inappropriate then you may well be giving him bad advice. So great care is needed here.

3. We assume in this analysis that the EMV criterion of maximizing the company's expected pay-off was an appropriate one to use. Unfortunately this is often not the case. In the next chapter we develop ways of defining an optimal decision when it is inappropriate to use the EMV criterion.

## EXERCISES

2.1 On his twentieth birthday a patient is brought into a hospital with an illness which is either Type I, with probability 0.4, or Type II, with probability 0.6. Independent of the type of illness, without treatment he will die on that day with probability 0.8 and otherwise survive and have normal life expectancy.

The surgeon has three possible courses of action open to him:

(i) not to treat the patient;
(ii) to give the patient drug $L$ once;
(iii) to operate on the patient once.

He cannot both operate and administer the drug.

Both operating and administering the drug are dangerous to the patient. Independently of the type of illness, operating on the patient will kill him with probability 0.5 and the drug will kill him with probability 0.2.

If the patient survives the poisonous effects of the drug, it will either cure him or have no effect, each with probability 0.5, if he has Type I illness; and will have no effect if he has Type II illness. If the patient survives an operation, it will cure him with probability 0.8 if he has Type I illness and with probability 0.4 if he has Type II illness, otherwise having no effect. Survival of the patient will give him a life expectancy of 70 years in all cases.

Draw a decision tree to represent this problem. Calculate the surgeon's best strategy assuming that he wishes to maximize his patient's life expectancy.

2.2 A company called 'Prune', which currently markets personal computers, will need to choose between at most three options in three years' time:

$a_1$ – continue to market its current machine (called Prunejuice);
$a_2$ – market an improved version of Prunejuice instead (called Pruneplus);
$a_3$ – market a much more powerful machine instead (called Superprune)

Prune can choose between $a_1, a_2, a_3$ but cannot implement more than one of them.

A machine can be marketed only if it has been researched and developed (R&D) successfully. The event that an R&D programme could be successful for Pruneplus and the event that it would be successful for Superprune are considered independent with current respective probabilities 0.9 and 0.6. If neither of the new machines has been successfully researched and developed in the next three years then Prune would be forced to choose option $a_1$ – to market its current machine. Prune now needs to choose whether to:

(i) R&D neither machine (decision $d_1$) at a cost of \$0;
(ii) R&D Pruneplus only (decision $d_2$) at a cost of \$3 000 000;
(iii) R&D Superprune only (decision $d_3$) at a cost of \$5 000 000;
(iv) R&D both Pruneplus and Superprune (decision $d_4$) at a cost of \$8 000 000.

Given successful R&D, Prune expects to make \$2 000 000 net profit from Prunejuice, \$10 000 000 net profit from Pruneplus and \$18 000 000 net profit from Superprune. Draw a decision tree representing Prune's *current* decision problem, and advise Prune on its best plan of action given that they want to maximize their expected net profit less costs.

2.3 A customer insists you give him a guarantee that a piece of machinery will not be faulty for one year. As the supplier you have the option of overhauling your machinery before delivering it to the customer (action $a_2$) or not (action $a_1$). The pay-offs (in \$) you will receive by taking these actions when the machine is, or is not, faulty are given below.

| | Action | |
|---|---|---|
| State | $a_1$ | $a_2$ |
| Not faulty | 1000 | 800 |
| Faulty | 0 | 700 |

The machinery will work if a certain plate of metal is flat enough. You have scanning devices which sound an alarm if a small region of this plate is bumpy. The probabilities that any one of these scanning devices sound an alarm given that the machine is, or is not faulty, are respectively 0.9 and 0.4. You may decide to scan with $n$ scanning devices ($d_n$), $n = 0, 1, 2, 3, \ldots$, at the cost of \$50$n$ and overhaul the machine or not depending on the number of alarms that ring. These devices will give independent readings conditional on whether the machine is faulty or not. Prior to scanning you believe that the probability that your machine is faulty is 0.2. Use the expected value of perfect information to determine those decision rules which might be optimal. Draw a decision tree of these decision sequences to find the optimal number of scanning devices to use and how to use them to maximize your expected pay-off.

# 3

# Utilities and rewards

## 3.1 INTRODUCTION

In the previous two chapters we assumed it was appropriate for a decision-maker to use the EMV criterion to find his best decision. Unfortunately real life is not so simple. It is often quite inappropriate to use the EMV algorithm.

*Example 3.1*

You are given a choice between two decisions: decision $d_1$, to take \$20 000 unconditionally, and decision $d_2$, to take \$40 000 if the toss of a fair coin resulted in heads and take nothing if the toss resulted in tails.

Which of these two options do you prefer? I think most people would prefer $d_1$ to $d_2$. On the other hand, if you need to raise \$40 000 by tomorrow (for example to pay off a creditor or risk losing your business) then you might prefer $d_2$ to $d_1$. In either case it is not at all clear that it is sensible to consider the two decisions as equivalent *regardless* of your circumstances. And this is exactly what is implicitly assumed by employing the EMV algorithm, since the expected pay-off from each decision is the same.

Of course a proponent of the EMV algorithm might suggest that the fact that you do not consider these two decisions as equivalent is just evidence of your own irrationality. The next example should help to convince you that there are games of chance where employing the EMV algorithm chooses as 'optimal' a quite ridiculous decision.

*Example 3.2 A game illustrating the 'St Petersburg paradox'*

A casino makes repeated independent tosses of a fair coin until a tail occurs. A gambler, starting with a stake of \$1, is offered the following wager.

After each toss the gambler will be given two choices. He may either take away his winnings from the previous tosses of the coin. In this case the game will end. Alternatively he may use all his winnings from previous tosses plus his original stake money as a stake for the next toss of the coin. This stake will be tripled by the casino if a head is tossed on the next throw. On the other hand, if a tail is thrown the gambler will lose all his winnings from previous tosses of the coin together with his original stake money.

Suppose that the gambler is instructed to follow the EMV algorithm when playing this game. Assume $r$ consecutive heads have been thrown and denote the gambler's original stake plus total winnings as $S_r$. His expected pay-off for withdrawing from the game is clearly $S_r$. However, his expected pay-off for continuing to play is at least $(\frac{1}{2})3S_r + (\frac{1}{2})0 = (\frac{3}{2})S_r$ (the expected pay-off for playing once more). So under the EMV algorithm the gambler should continue to stake his winnings until a tail is thrown. But since a tail will be thrown eventually with probability one, by following the EMV algorithm the gambler ensures that he will lose his original stake money *with certainty*!

Clearly, in the simple game given above, very rational people will not want to follow the dictates of the EMV algorithm. It is therefore necessary to modify the EMV algorithm so that 'optimal decisions' can be defined sensibly for situations like the one given above. It will be shown in the next section that such a modification is possible provided that your client is prepared to commit himself to following certain rules (or axioms). It also generalizes the EMV approach to problems when the client's objectives are not only the maximization of pay-off. Our first step is to introduce the idea of a reward.

The development given here parallels that of DeGroot (1970).

## 3.2 A SET OF RULES DEFINING SENSIBLE PREFERENCES

One of the first things you need to do when tackling a complicated decision problem is to encourage your client to be specific about what he hopes to achieve through the analysis. Typically he will have several objectives in mind that he hopes to achieve simultaneously, at least partially. For example, a businessman may on the one hand be interested in maintaining his cashflow, requiring actions that ensure at least a minimum amount of short-term cash (the larger the better). On the other hand, he must also keep an eye on the long-term viability of his firm and so be rewarded by the long-term pay-off of any action he takes. Thus by pursuing any particular course of action he will achieve a reward (or penalty) reflecting that action's effect on cash flow *and* its effect on the long-term viability of the firm.

It is common for these 'rewards' to be a combination of quite different components. The National Health Service needs to balance the 'reward' of success of a treatment against the 'reward' for the cheapness of that treatment. Identifying which objectives the client hopes to achieve is synonymous with identifying which aspects of the future scenario he regards as rewarding him. The term 'reward' can be defined precisely by the following rule, provided we can assume that our client can be encouraged to think probabilistically.

*Rule 1*

If two decision rules $d_1$ and $d_2$ give rise to identical probability distributions of rewards, then $d_1$ and $d_2$ should be considered by the client as equally preferable.

This rule demands that your constructed sample space of rewards, called the *reward space*, be comprehensive enough to reflect all the objectives that govern your client's preferences. It is very easy to mis-specify a reward space. The following hypothetical example illustrates the difficulty.

*Example 3.3*

Your client is a doctor who needs to treat a fixed number of patients who have a disease which will kill them if they are not treated. However, the treatments he uses may either themselves be lethal or completely successful. The doctor tells you that he is only interested in the effect each possible treatment has on the probability distribution of the number of survivors $r$ of treatment and disease. So $r$ measures his reward. Can you believe your client?

**Answer** No, you cannot believe him. Consider the simple case when your client will only treat two patients $M_1$ and $M_2$. Let $A_i$ and $\bar{A}_i$ denote the event that $M_i$ dies or survives respectively, $i = 1, 2$. Suppose under treatment $d_1$,

$$P(A_1 \cap A_2) = 1/9, \quad P(A_1 \cap \bar{A}_2) = 2/9, \quad P(\bar{A}_1 \cap A_2) = 2/9, \quad P(\bar{A}_1 \cap \bar{A}_2) = 4/9.$$

Under treatment $d_2$,

$$P(A_1 \cap A_2) = 1/9, \quad P(A_1 \cap \bar{A}_2) = 0, \quad P(\bar{A}_1 \cap A_2) = 4/9, \quad P(\bar{A}_1 \cap \bar{A}_2) = 4/9.$$

The reward distributions $P_j$ associated with $d_j$, $j = 1, 2$ are identical, each being given by

$$P_j(r) = \begin{cases} 0 \text{ with probability } 1/9 \\ 1 \text{ with probability } 4/9 \\ 2 \text{ with probability } 4/9 \end{cases}$$

However, surely this doctor must prefer $d_1$ to $d_2$ on ethical grounds, because whereas the first treatment ensures that the probability of survival for either patient is the same, under the second $M_1$ survives with high probability while $M_2$ almost certainly dies. Thus under treatment $d_2$ the second patient will be used as a 'guineapig' to enhance the first patient's chances of survival.

One of the most common errors made by the decision analyst is not to listen carefully enough to the client and produce an answer to his problem which is optimal only under an inappropriate set of objectives. Rule 1 demands that our chosen reward space is large enough to capture all the preferences of a client.

**Notation** To simplify some of the statements about preferences it is expedient at this point to introduce some notation.

Write $P_1 \overset{*}{<} P_2$ if your client prefers a decision giving rise to a distribution of rewards $P_2$ to a decision giving rise to distributions of rewards $P_1$.

Write $P_1 \overset{*}{=} P_2$ if decisions giving rise to distributions of rewards $P_1$ and $P_2$ respectively, are equally preferred.

In Section 3.3 we shall consider degenerate distributions of reward, called $r$, which assign a reward $r$ with probability one. The notation above can be used in the obvious way to denote preferences among rewards. For example, $r_2 \overset{*}{>} r_1$ reads 'reward $r_2$ is preferred to reward $r_1$'.

Any decision rule will determine a probability distribution of rewards. From Rule 1, two decisions which give the same distributions over rewards are considered equally preferable. Suppose you know your client's preferences over *distributions* of rewards. You can then deduce which of the *decision rules* offered to your client are most preferable.

Because of this we can now focus our attention on the client's preferences over distributions of rewards. We assume that preferences exist across all distributions of reward and satisfy certain 'logical' constraints. The first of these rules is given below. Let $\mathbb{P}$ denote the class of all distributions of rewards.

*Rule 2 The total ordering of hypothetical distributions of rewards*

(i) *Comparability* For all $P_1, P_2 \in \mathbb{P}$, either $P_1 \overset{*}{<} P_2$, $P_1 \overset{*}{=} P_2$ or $P_1 \overset{*}{>} P_2$.

(ii) *Transitivity* For any three distributions $P_1, P_2, P_3 \in \mathbb{P}$,

$$\begin{aligned} P_1 \overset{*}{<} P_2 &\text{ and } P_2 \overset{*}{<} P_3 \Rightarrow P_1 \overset{*}{<} P_3 \\ P_1 \overset{*}{<} P_2 &\text{ and } P_2 \overset{*}{=} P_3 \Rightarrow P_1 \overset{*}{<} P_3 \\ P_1 \overset{*}{=} P_2 &\text{ and } P_2 \overset{*}{<} P_3 \Rightarrow P_1 \overset{*}{<} P_3 \\ P_1 \overset{*}{=} P_2 &\text{ and } P_2 \overset{*}{=} P_3 \Rightarrow P_1 \overset{*}{=} P_3 \end{aligned}$$

Rule 2(i) implies that your client can state preferences between any two distributions of rewards, whether hypothetical or not. Rule 2(ii) implies that these preferences cohere in the natural way. For example, if your client prefers $P_2$ to $P_1$ and $P_3$ to $P_2$ you can deduce that he also prefers $P_3$ to $P_1$.

Paradoxically, Rule 2 complicates a decision problem by embedding the client's original decision space into one which is much larger. In particular, the client may be asked to choose between many simple but hypothetical reward distributions associated with decision rules it may not be possible to offer him in practice. However, by studying his preferences over these simple decisions and assuming that his preferences are related in a 'rational' way (rational being defined by Rules 2, 3 and 4) you are able to deduce what his preferences should be over all decision rules, no matter how complicated the corresponding reward distributions might be. In particular, it is possible to discuss which decision is best for him given that he must choose his decision/reward distribution from a given set of options.

The next two rules define further structure on what might be considered a sensible total ordering on distributions of rewards.

*Rule 3 The consistent ordering of lotteries*

Suppose $P_1$, $P_2$ and $P$ are any three distributions of reward. Then for all probabilities $0 < \alpha < 1$,

$$P_1 \overset{*}{<} P_2 \text{ if and only if } \alpha P_1 + (1-\alpha)P \overset{*}{<} \alpha P_2 + (1-\alpha)P$$

where $\alpha Q_1 + (1-\alpha)Q_2$ denotes the reward distribution which assigns rewards using distribution $Q_1$ with probability $\alpha$ and $Q_2$ with probability $1-\alpha$.

*Example 3.4*

Let

$$P_2 = \begin{cases} 2000 & \text{with probability } \frac{1}{2} \\ 0 & \text{with probability } \frac{1}{2} \end{cases} \qquad \begin{aligned} P_1 &= 500 \text{ with probability } 1 \\ P &= 200 \text{ with probability } 1 \end{aligned}$$

$$P_3 = \begin{cases} 2000 & \text{with probability } \frac{1}{4} \\ 200 & \text{with probability } \frac{1}{2} \\ 0 & \text{with probability } \frac{1}{4} \end{cases} \qquad P_4 = \begin{cases} 500 & \text{with probability } \frac{1}{2} \\ 200 & \text{with probability } \frac{1}{2} \end{cases}$$

$$= \alpha P_2 + (1-\alpha)P, \ \alpha = \tfrac{1}{2} \qquad\qquad = \alpha P_1 + (1-\alpha)P, \quad \alpha = \tfrac{1}{2}$$

If a client's rewards can be measured solely in terms of monetary pay-off, then Rule 3, with $\alpha = \frac{1}{2}$, implies in particular that $P_2$ is preferred to $P_1$ if and only if $P_3$ is preferred to $P_4$.

Most rational clients should be prepared to satisfy Rule 3. To see this, suppose a client is given the hypothetical opportunity of entering the following lottery between two options $b_1$ and $b_2$. Under either option he will obtain a reward from a distribution $P$ with probability $1-\alpha$. However, with probability $\alpha$ he will be allocated a reward from the distribution $P_i$ if he chooses bet $b_i$, $i = 1, 2$. Surely, if he is acting sensibly he should prefer $b_2$ to $b_1$ if he prefers $P_2$ to $P_1$. Conversely if he prefers $b_2$ to $b_1$ surely you can infer from this that he prefers $P_2$ to $P_1$. This is precisely how Rule 3 requires a client to structure his preferences.

Our final rule is somewhat more contentious.

*Rule 4 The comparability of rewards*

Let $P_1$, $P_2$ and $P$ be any three distributions of reward such that $P_1 \overset{*}{<} P \overset{*}{<} P_2$. Then there exists values of $\alpha$ and $\beta$, $0 < \alpha, \beta < 1$, such that, under the notation of the previous rule,

(i) $P \overset{*}{<} \alpha P_2 + (1-\alpha)P_1$
(ii) $P \overset{*}{>} \beta P_2 + (1-\beta)P_1$

If a client will not accept Rule 4(i), then no matter how small is the chance of obtaining reward distribution $P_1$ he will not find a lottery between $P_1$

and $P_2$ preferable to the reward distribution $P$. Consequently he would believe that in this sense the reward distribution from $P_1$ was infinitely worse than that of either $P$ or $P_2$. In a similar sense, if he wishes to break Rule 4(ii) you can conclude that he believes $P_2$ to be an infinitely better reward than either $P$ or $P_1$.

There are sometimes strong cases to be made for a client to choose to break Rule 4, especially when his reward space has more than one dimension. Here is an example to illustrate this point.

*Example 3.5*

A doctor's reward space is pair $(x, -y)$ where $x = 1$ if a patient survives a given treatment and $x = 0$ if she dies, and where $y$ is the cost of a treatment. A cheap treatment $T_2$, costing $\$y_2$, is always successful, i.e. $x = 1$. So under treatment $T_2$ the reward distribution $P_2 = (1, -y_2)$ with probability 1. A second treatment $T$ is a little more expensive (costing $\$y_1$) and is equally successful as $T_2$ in curing patients. $P = (1, -y_1)$ with probability 1 is therefore the reward distribution associated with $T$. A third treatment $T_1$ is equally as expensive as $T$ but only half the patients given this treatment survive. The reward distribution $P_1$ of $T_1$ is thus given by:

$$P_1 = \begin{cases} (0, -y_1) \text{ with probability } \frac{1}{2} \\ (1, -y_1) \text{ with probability } \frac{1}{2}. \end{cases}$$

Clearly any doctor would prefer $T_2$ to $T$ and $T$ to $T_1$ so that

$$P_1 \overset{*}{<} P \overset{*}{<} P_2$$

Your doctor is now given the option of a 'randomized' treatment $T(\alpha)$ which allocates treatment $T_2$ with probability $\alpha$ and $T_1$ with probability $(1 - \alpha)$. Because he believes that the survival of a patient is infinitely more important than the cost of treatment the doctor may, quite legitimately, always prefer $T$ to $T(\alpha)$, regardless of the value of $\alpha$, and hence break Rule 4.

In one sense Rule 4 is complementary to Rule 1. Remember that Rule 1 could be interpreted as requiring that the reward space you choose for a client to be large enough to encompass all his preferences. On the other hand, Rule 4 requires from your client that he use a reward space whose components are not so different in *priority* that they cannot be compared using lotteries of the above type.

## 3.3 THE CONSEQUENCES OF SENSIBLE PREFERENCES

Suppose that through a discussion of your client's decision problem it is clear that he is prepared to follow Rules 1–4. It can then be shown that for any three rewards $r_1 \overset{*}{<} r \overset{*}{<} r_2$, there exists a *unique* value of $\alpha$, $0 < \alpha < 1$, such that

$$r \overset{*}{=} \alpha r_2 + (1-\alpha)r_1 \tag{3.1}$$

The proof of this result is a little technical so I have relegated it to Appendix 1 (Lemma A1.1). This relates a reward $r$, which may be multidimensional, to a single real number $\alpha \in (0, 1)$. As a consequence it can be shown that there exist a real function $U$ of $r$ such that for any three rewards $r_1, r_2, r, r_1 < r < r_2$,

$$\begin{aligned} r &\overset{*}{=} \alpha r_2 + (1-\alpha)r_1 = P(\alpha) \\ u(r) &= \alpha u(r_2) + (1-\alpha)u(r_1) = \bar{u}(P(\alpha)) \end{aligned} \tag{3.2}$$

where

$$u(r_1) < u(r_2) \text{ wherever } r_1 \overset{*}{<} r_2 \tag{3.3}$$

Furthermore, from equation (3.1) it can be deduced that if $u^*$ is another function of $r$ satisfying equations (3.2) and (3.3) then

$$u^* = au + b \qquad a > 0 \tag{3.4}$$

where $a$ and $b$ do not depend on $r$. The proof of these results when there exists a most favourable and a least favourable reward is given in Lemma A1.3 of Appendix 1. (A proof which does not rely on the existence of a least and most favourable reward is given in DeGroot (1970).)

*Definition*

Any function $u$ of $r$ satisfying (3.2) and (3.3) is called a *utility function* of the client.

So if a client agrees to follow Rules 1–4 you can express his preference for a reward $r$ by a real index $u(r)$ of desirability. The larger the number $u(r)$, the more preferable is the reward $r$.

Now notice that for distributions $P(\alpha)$ over two rewards we have the property from equation (3.2) that

$$r \overset{*}{=} P(\alpha) \Rightarrow \bar{u}(r) = u(r) = \bar{u}(P(\alpha))$$

where $\bar{u}(P)$ denotes the expected utility given a distribution of rewards $P$. It follows that two distributions $P_1$ and $P_2$ each assigning probability zero to all but two points, are equally preferable only if their expected utilities are equal. It is not difficult to show that this result extends to all distributions over finite numbers of possible rewards and that the implication can be reversed. For a proof of these results see Theorem A1.1 of Appendix 1. This gives us the following remarkable theorem.

*Theorem 3.1 The utility theorem*

Suppose a client agrees to comply with Rules 1–4. In addition suppose that each possible decision $d_i$ gives rise to a reward distribution $P_i$ which assigns zero probability to all but a finite set $S(P_i)$ of rewards where this set may depend on the decision used. Then for all values $i, j$, $P_i \overset{*}{>} P_j$ if and only if

$$\bar{u}(P_i) > \bar{u}(P_j)$$

where $\bar{u}(P)$ denotes the expected utility of rewards given a distribution $P$.

## 3.4 USES OF THE UTILITY THEOREM

Suppose there exists a decision $d^*$ such that $\bar{u}(d^*) \geqslant \bar{u}(d)$ for all possible decision $d$, where $\bar{u}(d)$ denotes the expected utility on reward given decision $d$ is taken. Then $d^*$ is called a *Bayes decision under utility u*. A Bayes decision under utility $u$ exists for most problems and is guaranteed if the space of all possible decisions is compact† and $u$ is continuous. Clearly a Bayes decision is one which is most preferable to your client. Notice that the definition of $d^*$ is unambiguous in the sense that if $d^*$ maximizes $\bar{u}(d)$ it will also maximize $\bar{u}^*(d)$ where $u^*(d)$ is any function satisfying

$$u^*(d) = au(d) + b \qquad \text{where } a > 0$$

Finally notice that, because the EMV algorithm defines as optimal a decision which maximizes expected pay-off, this decision is a Bayes decision with respect to a linear utility on reward which is simply monetary pay-off.

To summarize: to find an optimal decision for a client with reward space $R$ you first find his utility over $R$ by considering simple lotteries of the form given in equation (3.2). Having fixed utility values to each reward in the reward space $R$ you then treat these values like monetary pay-off and follow in EMV strategy, i.e. find the decision which maximizes your expected utility rather than expected pay-off.

Now in fact the utility theorem given above can be extended to hold true for *all* distributions of reward, whether discrete or continuous, provided that a couple more rules can be complied with. These rules are technical and can be found in DeGroot (1970). They are not at all stringent rules; they ensure that various limits of preferences on rewards behave in a predictable way. Provided your client is prepared to accept these rules then problems like the St Petersburg paradox can be solved. Here is an example of how it is done.

*Example 3.6 The St Petersburg paradox revisited*

In Example 3.2 assume now that the gambler has a utility function on reward $r$ of the form

$$u(r) = r(e^{\theta} + r)^{-1} \qquad r > 0 \tag{3.5}$$

The parameter $\theta$ will reflect the gambler's propensity to take risks. The larger the value of $\theta$, the more he prefers speculative gains of large amounts to the

†See, for example, Fullerton (1971)

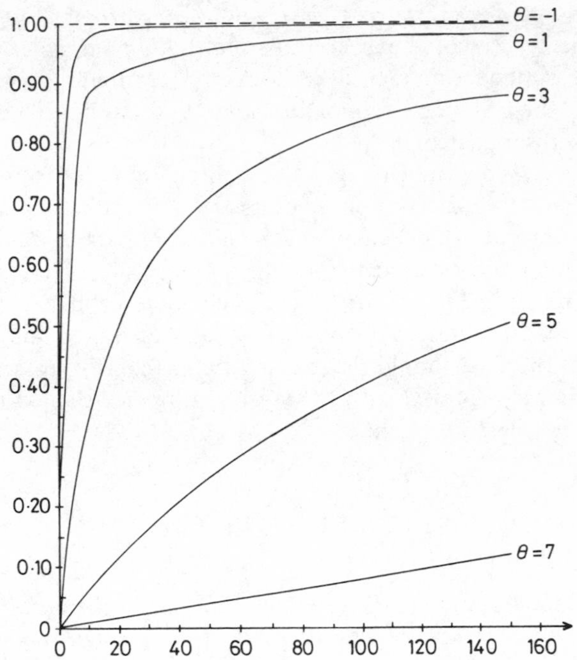


**Figure 3.1** Some examples of the utility curve $u(r) = r(e^{\theta} + r)^{-1}$

certainty of gaining small amounts. The utility $u(r)$ is shown in Fig. 3.1 for certain values of $\theta$. Regardless of $\theta$, $u(r)$ is concave and takes values between 0 and 1.

The decision-maker's expected utility associated with the decision $d_n$ of terminating the game after the $n$th toss is

$$\bar{u}(d_n) = (\tfrac{1}{2})^{n-1}3^{n-1}(e^{\theta} + 3^{n-1})^{-1} = (1.5)^{n-1}(e^{\theta} + 3^{n-1})^{-1} \tag{3.6}$$

It is easily checked that a function $e^{\mu x}[e^{\theta} + e^{\lambda x}]^{-1}$ is maximized when

$$x = \lambda^{-1}[\log \mu - \log(\lambda - \mu) + \theta] \qquad \text{if } \lambda > \mu$$

It follows that $(1.5)^{y-1}(e^{\theta} + 3^{y-1})^{-1}$ is maximized when

$$y = \lambda^{-1}[\log \mu - \log(\lambda - \mu) + \theta] + 1 \qquad \text{where } \mu = \log 1.5, \quad \lambda = \log 3$$

$\bar{u}(d_n)$ is therefore maximized at either the largest non-negative integer value $n$ less than $y$ or the smallest non-negative integer greater than $y$, where $y$ is defined above. Notice in particular that $y$ increases (linearly) with $\theta$. So the more ready the gambler is to risk large speculative gains, the longer he should play the game, as we would expect. But note that any gambler with a utility function given in equation (3.5) will choose to terminate the game at some stage – unlike the EMV decision-maker. So then the supposed St Petersburg 'paradox' has disappeared.

The utility curves depicted in Fig. 3.1 have been estimated from five different students in a class of 43. The parameters $\theta = -1$ and $\theta = 7$ are estimated values of the most extreme students in the class. The Bayes decisions for the time $d^*$ of termination of the game for $\theta = -1, 1, 3, 5, 7$ are respectively found to be $0, 0, 1, 3$, and 5. In practice these curves were estimated very crudely. However, if we later found that they were accurate we could conclude that no member of the class should risk buying more than five tosses of the coin. This conclusion agreed well with the stated strategies elicited directly from members of this class

## 3.5 FINDING A UTILITY FUNCTION WHEN REWARDS ARE MONETARY

We now turn our attention to the problem of finding a client's utility function. This is easiest if your client's reward happens to be a single monetary pay-off. In theory all you would need to do is to use equation (3.2) in the following way. Begin by choosing two arbitrary pay-offs $s$ and $t$ with $s < t$ and set $u(s) = 0$ and $u(t) = 1$.

**Case I** If pay-off $r > t$, by equation (3.1) there is a unique value of $\alpha$ such that

$$t \overset{*}{=} \alpha r + (1 - \alpha)s$$

and by equation (3.2),

$$u(t) = \alpha u(r) + (1 - \alpha)u(s)$$

which, since $u(s) = 0$ and $u(t) = 1$, implies $u(r)$ is uniquely defined by

$$u(r) = 1/\alpha$$

To find $u(r)$ in this case all you need to do is to find that value of the probability $\alpha$ at which your client is indifferent between obtaining pay-off $t$ with certainty and obtaining pay-off $r$ with probability $\alpha$ and pay-off $s$ with probability $1 - \alpha$.

**Case II** If pay-off $s < r < t$, then using equations (3.1) and (3.2) when $\alpha$, $0 < \alpha < 1$, is such that

$$r \overset{*}{=} \alpha t + (1 - \alpha)s,$$

$$u(r) = \alpha u(t) + (1 - \alpha)u(s) = \alpha$$

**Case III** If pay-off $r < s$, then using equations (3.1) and (3.2) when $\alpha$, $0 < \alpha < 1$, is such that

$$s \overset{*}{=} \alpha t + (1 - \alpha) r,$$
$$u(s) = \alpha u(t) + (1 - \alpha) u(r)$$

i.e.

$$u(r) = -\alpha(1 - \alpha)^{-1}$$

In both these cases $u(r)$ can be evaluated by finding the value of $\alpha$ giving rise to the required indifference.

In practice, if you try to measure utility in the way described above, then the answers you obtain will be subject to large measurement errors. They are particularly poor in this respect if the corresponding value of $\alpha$ is very close to either 0 or 1. There are various ways of improving this measurement (see Keeney and Raiffa, 1976, Ch. 4). A simple one is described below.

Let $s(0)$ denote the smallest possible and $s(1)$ the largest possible pay-off. Let $u(s(0)) = 0$ and $u(s(1)) = 1$. Now find the pay-off $s(\frac{1}{2})$ for which your client is indifferent between obtaining $s(\frac{1}{2})$ with certainty and obtaining $s(0)$ if a head occurs on the toss of a fair coin and $s(1)$ otherwise. It will follow from equation (3.2) that $u(s(\frac{1}{2})) = \frac{1}{2}$. Next find the pay-off $s(\frac{1}{4})$ for which your client is indifferent between obtaining $s(\frac{1}{4})$ with certainty and obtaining either $s(0)$ or $s(\frac{1}{2})$ depending on the outcome of the toss of a fair coin. Let $s(\frac{3}{4})$ be the pay-off your client is indifferent to compared with a fair toss between pay-off $s(\frac{1}{2})$ and $s(1)$. It is easily checked from equation (3.2) that $u(s(\frac{1}{4})) = \frac{1}{4}$ and $u(s(\frac{3}{4})) = \frac{3}{4}$. Continuing in this way you can obtain all those points $u[s(k/2^n)] = k/2^n$, $k = 0, 1, \ldots, 2^n$, for some large value of $n$. For most practical purposes a piecewise linear curve through these pay-offs $s(k/2^n)$ should now be a sufficiently close approximation to the client's true utility function.

One point you will find when measuring client's utility functions is that in practice $u$ is nearly always concave – that is, if $r \overset{*}{=} \frac{1}{2}t + \frac{1}{2}s$, then $r \leqslant \frac{1}{2}(t + s)$. A decision-maker having such a utility is called *risk-averse*. All the utility functions in Example 3.6 are concave. It is because people are often naturally strictly risk-averse that the EMV 'optimal' solution to the betting game of Example 3.2 seems so perverse.

Having found a client's utility function, the techniques discussed so far for obtaining optimal solutions to problems using the EMV algorithm can be adapted very simply to find decisions which maximize utility. Here is what you do when you have drawn a decision tree.

*Example 3.7 Finding a Bayes decision under a given utility from a decision tree*

Suppose you want to find your client's Bayes decision given his utility having expressed his problem as a decision tree. Having drawn his decision tree and filled in the pay-offs to the tips of each branch, replace these pay-offs with their utility value. Now proceed exactly as before where expected utilities replace expected pay-off. At the trunk of the tree you will finish with each of the initial branches having associated expected utilities. Your Bayes decision rule then just corresponds to the initial branch with the largest expected utility.

Two words of warning are given here. First, note that the expected utility associated with the Bayes decision is *not* the expected pay-off. You can work back and calculate the pay-off distribution associated with the Bayes decision rule but this typically requires extra computation.

Secondly, terminal pay-offs associated with pairs of decisions and outcomes are in fact in practice often *expected* pay-offs. Before you can proceed with the methodology given above you need extra information about the terminal pay-off distribution so as to calculate expected utilities. It may be very difficult to access this information.

## 3.6 WHAT TO DO WHEN A CLIENT'S REWARDS HAVE MORE THAN ONE COMPONENT

As we have seen, for many interesting problems a client's reward space is more than one-dimensional. A businessman has to balance immediate pay-off arising from his decision against its implications for the future. A doctor needs to balance the effectiveness of a given treatment against its cost. When there is more than one component in a client's reward space we are solving a *multi-attribute decision problem.*

In such a problem it would be possible, though laborious, to assess $u(r)$ by the methods described in the last section, using equations (3.1) and (3.2) with your 'multi-attribute' rewards replacing your pay-offs. However, when a client is asked to balance simultaneously the pros and cons of options with respect to many different objectives he will tend to get confused. His stated preferences will then often be unreliable. To simplify the elicitation of a client's utility function it is very helpful to be able to assume that his preferences over distributions of rewards satisfy a further rule.

Let the client's reward $\mathbf{r} = (r(1), r(2), \ldots, r(n))$. The components $r(i)$ of $\mathbf{r}$, $1 \leqslant i \leqslant n$, are called its *attributes.* For example, if $n = 2$, $r(1)$ might label the cost of a project this year and $r(2)$ the cost of that project next year. Together, these two attributes might determine the client's reward.

In this example, the decision analyst can ask the client not to concern himself with the consequences of any policy on next year's costs but to concern himself only with assessing his preferences over policies as they affect costs this year. Such preferences are called his *marginal preferences* for attribute $r(1)$. By the utility theorem these marginal preferences will be determined through his marginal distributions of attribute $r(1)$ arising from each policy he considers. Suppose his marginal preferences on $r(1)$ say that policy $d_1$ is equally preferred to $d_2$. Suppose also that this marginal preferences on $r(2)$, elicited analogously, also state that policy $d_1$ is equally preferred to $d_2$. What we require for the next rule to hold is that if $d_1$ and $d_2$ are equally preferable marginally over the two attributes they are equally preferable to the client in the original problem. Generalizing this idea to $n$ attributes we thus have the following rule.

*Rule 5. The value independence of attributes*

In the client's problem the attributes $r(i)$, $1 \leqslant i \leqslant n$, are value independent. (Attributes are said to be *value independent* if two distributions on reward $\mathbf{r}$ are equally preferred whenever the client is indifferent between the two distributions on the corresponding marginal preferences on each attribute $r(i)$, $1 \leqslant i \leqslant n$.)

Rule 5 is important because it leads us to the following theorem.

*Theorem 3.2*

Suppose a client's preferences satisfy Rule 5. Then for every reward $\mathbf{r}$, the client's utility $u(\mathbf{r})$ can be written in the form

$$u(\mathbf{r}) = \sum_{i=1}^{n} a_i u_i(r(i)) \tag{3.7}$$

where $a_i > 0$, $1 \leqslant i \leqslant n$, are real values not depending on $\mathbf{r}$, and $u_i$ is a function of the component $r(i)$ only, $1 \leqslant i \leqslant n$.

A proof of this theorem can be found in Keeney and Raiffa (1976).

Marginal preferences are usually difficult to measure accurately and so it is not recommended that the client's utility function be elicited directly using Rule 5 and (3.7). However, provided a client is happy to follow Rule 5, we can use the form of utility given in equation (3.7) to enable us to construct a client's utility function by comparing the desirability of rewards that only differ in one of the components $r(i)$ of $\mathbf{r}$. I shall illustrate below how this is done. To simplify the argument I will assume that, for each component of the rewards $r(i)$, there is a least preferable and most preferable value denoted respectively by $s(i)$ and $t(i)$.

First set $u_i(s(i)) = 0$ and $u_i(t(i)) = 1$, $1 \leqslant i \leqslant n$. This can be done without loss of generality (see Exercise 3.8). We now compare three rewards $\mathbf{r}_1, \mathbf{r}_2, \mathbf{r}$ whose $n-1$ components other than $r(i)$ are fixed to some preassigned value and where the $i$th component of $\mathbf{r}_1$ is the least desirable reward $s(i)$ and where the $i$th component of $\mathbf{r}_2$ is the most desirable reward $t(i)$. As a consequence of Rules 1–4, for the (unique) value of $\alpha$ for which $\mathbf{r} \overset{*}{=} \alpha\mathbf{r}_2 + (1-\alpha)\mathbf{r}_1$,

$$u(\mathbf{r}) = \alpha u(\mathbf{r}_2) + (1-\alpha)u(\mathbf{r}_1) \tag{3.8}$$

But, assuming Rule 5,

$$u(\mathbf{r}_1) = \sum_{j=1}^{n} a_j u_j(r(j))$$

$$u(\mathbf{r}_1) = a_i u_i(s(i)) + \sum_{j \neq i = 1}^{n} a_j u_j(r(j))$$

$$u(\mathbf{r}) = a_i u_i(r(i)) + \sum_{j \neq i = 1}^{n} a_j u_j(r(j))$$

$$u(\mathbf{r}_2) = a_i u_i(t(i)) + \sum_{j \neq i = 1}^{n} a_j u_j(r(j))$$

Substituting these values into (3.8) and remembering that $u_i(s(i)) = 0$ and $u_i(t(i)) = 1$ gives us that

$$u_i(r(i)) = \alpha \tag{3.9}$$

To identify a utility function $u$ for a reward $\mathbf{r} = (r(1), r(2), \ldots, r(n))$ from equation (3.7) we need only identify suitable values of $a_1, \ldots, a_n$ appearing in this equation.

Let
$$\mathbf{s} = (s(1), s(2), \ldots, s(n))$$
$$\mathbf{t} = (t(1), t(2), \ldots, t(n))$$

where $s(i)$ and $t(i)$ are defined above, $1 \leqslant i \leqslant n$, and let $\mathbf{t}_i$ be defined by

$$\mathbf{t}_i = (t_i(1), t_i(2), \ldots, t_i(n)) \qquad \text{where } t_i(i) = t(i),$$
$$t_i(j) = s(j),\ 1 \leqslant j \neq i \leqslant n.$$

As a consequence of Rules 1–4, for the (unique) value of $\alpha_i$ for which $\mathbf{t}_i \overset{*}{=} \alpha_i \mathbf{t} + (1 - \alpha_i)\mathbf{s}$,

$$u(\mathbf{t}_i) = \alpha_i u(\mathbf{t}) + (1 - \alpha_i) u(\mathbf{s}), \qquad 1 \leqslant i \leqslant n \tag{3.10}$$

Since $u_i(s(i)) = 0$ and $u_i(t(i)) = 1$, $1 \leqslant i \leqslant n$, by equation (3.7)

$$u(\mathbf{t}_i) = a_i$$
$$u(\mathbf{t}) = 1 \qquad \text{provided } \sum_{i=1}^{n} a_i = 1.$$
$$u(\mathbf{s}) = 0$$

It follows from equation (3.10) that

$$a_i = \alpha_i \qquad \text{provided } \sum_{i=1}^{n} \alpha_i = 1.$$

So comparing preferences using betting schemes defined in (3.8) and (3.10) each of which only consider betting schemes differing in one component of $\mathbf{r}_i$ it is possible, using Rule 5, to construct your client's utility function over any vector of rewards.

In practice, the method described above may well not be the most accurate way of eliciting $u$. A comprehensive account of how you should proceed is given in Keeney and Raiffa (1976).

Although it is often expedient to use Rule 5 it is difficult to determine whether the client's utility is of the right form. This is because to check Rule 5 properly you need the client's marginal preferences which are difficult to elicit because they are very hypothetical. You therefore need to check Rule 5 empirically if you are to have any confidence in the results it gives. One way to discover whether your client's utility is not in the required form is to find $\alpha$ repeatedly using the betting schemes (3.8) for several different fixed values of the components of $\mathbf{r}$ other than $r(i)$. Under the Rule 5 hypothesis you should obtain the same value of $\alpha$ whatever the values of the components of $\mathbf{r}$. Therefore if these values of $\alpha$ are markedly different you know that you cannot use Rule 5.

If you find Rule 5 breaks down in this way it may be that the components of reward are logically related in some way and that by reparametrizing them you can assume Rule 5 holds, at least approximately. If this does not work then there may exist a different preference rule which determines a different form of utility function. Some of these are outlined in Keeney and Raiffa (1976) and Bell, Keeney and Raiffa (1977). If all else fails you will be forced to rely on the laborious method of eliciting $u(\mathbf{r})$ given at the beginning of this section.

Multiattribute problems are very common and we have only very briefly sketched their analysis here. For more detailed discussions of case studies that give rise to multiattribute rewards see Keeney and Raiffa (1976), Phillips (1979) and Humphreys (1977).

## 3.7 SOME CONCLUDING REMARKS

We saw two examples to show that the EMV algorithm does not always produce sensible 'optimal decisions' and that for general decision problems a more foolproof method of identifying a best course of action was required. To solve your problems you needed to introduce the preferences of your client across the various outcome 'rewards' that might occur. By insisting that your client follows certain rules that ensure his preferences were coherent with one another and by determining his preferences in simple hypothetical problems you could isolate his utility function and hence the decision that was best for him given his current uncertainty and preferences.

Notice that by pursuing such algorithms we have abandoned the apparent objectivity of the EMV algorithm. However, we have gained something far more valuable – a strategy tailored to the practical needs of our clients.

In all the examples I have given so far I have assumed that the probabilities given in the problem are known. In practice at least some of the probabilities you are given will be numbers which reflect the client's own uncertainty about events rather than known logical quantities. Thus not only does the client influence the decision-making in terms of his preferences (utilities) but also in terms of his subjective probability assessments. These latter assessments will form the subject matter of the next chapter.

## EXERCISES

3.1 If you choose to play a game, decision $d_1$, you will be given a stake of \$1. A fair coin will then be tossed until a tail results when the game will end. Your stake money will have been doubled for each head that occurred before the first tail. You are also offered a decision $d(A)$ of being given an amount \$$A$ with certainty. Under an EMV strategy how large would $A$ have to be before $d(A)$ is preferred to $d_1$?

If your utility function $u(r)$ on pay-off $r$ is proportional to $\log_e r$, $r \geqslant 1$, show that you would prefer to play the game than receiving any amount less than 2.

3.2 (*Due to Professor Allias*) This is a criticism of the consequence of the utility axioms. Consider the two wagers depicted in Fig. 3.2.

Pick what you would consider your best decision for each of the two problems. Now suppose $\$1\,000\,000 \overset{*}{=} \alpha\$5\,000\,000 + (1-\alpha)\$0$. Show that, using the utility assumptions, we should never prefer $d_1$ to $d_2$ in Problem 1 and $d_3$ to $d_4$ in Problem 2 (or $d_2$ to $d_1$ and $d_4$ to $d_3$).

3.3 Decisions $d_1, d_2, \ldots, d_n$ are evaluated across a 3-dimensional reward space $\mathbf{r}(d) = (r_1(d), r_2(d), r_3(d))$. Suppose you choose to prefer $d_i$ to $d_j$ if and only if $r_k(d_i) > r_k(d_j)$ for at least two values of $k$, $k = 1, 2, 3$. By constructing an example, show that this strategy breaks Rule 2(ii).

3.4 Three investors $I_1, I_2, I_3$ each wish to invest \$10 000 in shareholdings. Four shares $S_1$, $S_2$, $S_3$, $S_4$ are available to each person. $S_1$ will guarantee the shareholder exactly 8% interest over the next year. Share $S_2$ will pay nothing with probability 0.1, 8% interest with probability 0.5, and 16% with probability 0.4. Share $S_3$ will pay 5% interest with probability 0.2 and

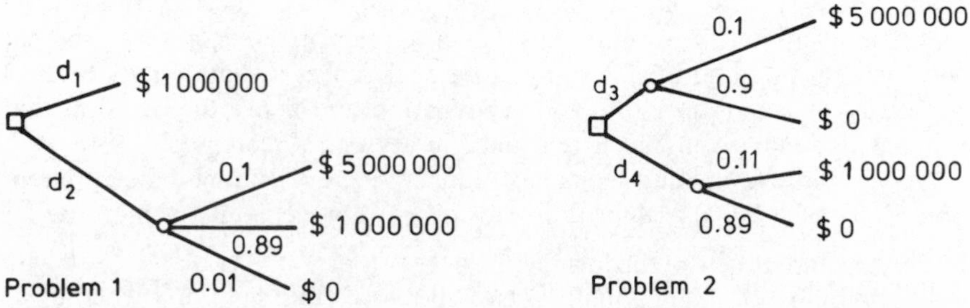


Figure 3.2

12% interest with probability 0.8. Share $S_4$ will pay 16% interest with probability 0.8 and nothing with probability 0.2 over the next year.

Each investor must list portfolios $P_1$, $P_2$, $P_3$, $P_4$, $P_5$ in order of preference. Portfolio $P_i$ represents the investment of \$10 000 in share $S_i$, $i = 1, 2, 3, 4$. Portfolio $P_5$ represents the investment of \$5000 in share $S_1$ and \$5000 in share $S_4$.

Investor $I_1$ prefers $P_1$ and $P_3$ to $P_5$. Investor $I_2$ prefers $P_1$ to $P_3$ and $P_3$ to $P_4$. Investor $I_3$ prefers $P_1$ to $P_4$ and $P_2$ to $P_1$.

Find the pay-off distribution over the next year associated with each portfolio. Hence, or otherwise, identify those investors whose preferences given above violate Rules 1–4 and explain why. You may assume that each investor's utility is an increasing function of his pay-off in the next year.

3.5 The table below gives the pay-offs in \$ associated with six possible decisions $d_1, \ldots, d_6$ where the outcome $\theta$ can take one of two possible values $\theta_1$ or $\theta_2$.

| | Decision | | | | | |
|---|---|---|---|---|---|---|
| Outcome | $d_1$ | $d_2$ | $d_3$ | $d_4$ | $d_5$ | $d_6$ |
| $\theta_1$ | 2 | 8 | 3 | 7 | 3 | 2 |
| $\theta_2$ | 2 | 2 | 3 | 1 | 4 | 6 |

(i) Find those decisions which can be Bayes decisions under a linear utility function for some values of the probability $p(\theta_1)$. Specify the corresponding possible values of $p(\theta_1)$ for each of these decisions.

(ii) Explain carefully which of the decisions $d_1, \ldots, d_6$ can be Bayes decisions under some utility function (assumed strictly increasing in pay-off) and probability value $0 < p(\theta) < 1$.

3.6 A customer will buy in bulk one of $n$ types of machine $M_1, \ldots, M_n$ that you manufacture, but has yet to commit himself to which machine it will be. Your probability that he will pick machine $M_i$ is $P_i, 1 \leqslant i \leqslant n$. You have a budget of \$$T$ to spend on the development of the different machines. The amount $r_i$ you spend on a machine $M_i, 1 \leqslant i \leqslant n$ will give rise to sales of $M_i$ proportional to $\log r_i$ provided the customer picks $M_i$, $1 \leqslant i \leqslant n$. What allocation of grants will maximize the expected sales of your product? (*Hint*: You may assume that:

$$\sum P_i \log(x_i) < \log(\sum P_i x_i) \qquad x_i, P_i > 0, 1 < i < n, \text{ and } \sum P_i = 1.$$

3.7 The *constant-risk decision-maker*. Suppose a decision-maker's preferences are such that bets of the form win $x + h$ with probability $\alpha$ and $x - h$ with probability $1 - \alpha$ are equally preferred to a certain gain of $x$ when $\alpha = p$ where $p$ does not depend on $x$, $0 < p < 1$. Show that his utility function $u(x)$

is either linear (when $p=\frac{1}{2}$), an increasing linear transformation of $e^{cx}$, $c>0$ (when $p<\frac{1}{2}$), or an increasing linear transformation of $1-e^{-cx}$, $c>0$ (when $p>\frac{1}{2}$). The last case defines the *constant risk-averse utility functions*.

3.8 In Section 3.6 we fixed $u_i(s(i))=0$ and $u_i(t(i))=1$, $1\leqslant i\leqslant n$. Show that by fixing $u_i(s(i))=L_i$ and $u_i(t(i))=M_i$, $L_i\leqslant M_i$, $1\leqslant i\leqslant n$, where $L_i$ and $M_i$ are arbitrary and by otherwise performing the same experiments on rewards differing in one component, you would derive a utility function which is an increasing linear transformation of the one derived in Section 3.6.

3.9 Most degree students choose a prearranged number of courses from a fixed set of options. You also may have many objectives in choosing your courses, e.g.:

(i) to obtain as high a mark as possible;
(ii) to need to work as little as possible to obtain reasonable marks;
(iii) to do an interesting course;
(iv) to find the course useful for your later career, etc.

Assuming Rule 5 is appropriate (see Section 3.6) list your objectives. For each possible course $d_i$ you might have taken, assign a utility value $u_j$ associated with the $j$th objective in the way described in this chapter. By using different fixed points, check the appropriateness of Rule 5 for this problem. What arguments are there against choosing to do the $n$ courses with the highest utility values when $n$ is the number of courses you must attend?

# 4

# Subjective probabilities and their measurement

## 4.1. AN INTRODUCTION TO SUBJECTIVE PROBABILITY

In all the examples we have discussed so far we have assumed that any probability values that are required in order to identify the client's best course of action are specified before the process of decision analysis takes place. In practice, of course, this is hardly ever the case. On the contrary, the probability assigned to an event will necessarily be a number between zero and one which summarizes the client's personal beliefs. In particular, measuring these values will be part of the decision analysis. Probabilities as measures of a client's beliefs are called *subjective probability assessments* and will be the subject of this chapter.

There has been considerable controversy over whether a client's beliefs should be summarized by a single probability value assigned to each event of interest, when it is assumed that these probabilities satisfy the usual axioms defining a probability measure over an event space.

Before addressing this controversy I will first outline two methods of quantifying a client's subjective probability of an event on the assumption that it *is* sensible to make such an assignment.

## 4.2 THE RELATIVE FREQUENCY APPROACH TO ASSESSING PROBABILITIES

Perhaps the simplest way of assigning a probability to a client's beliefs is to use the *relative frequency approach*. Here the client's subjective probability is related to the proportion of successes in independent trials as illustrated below.

Refer to Example 1.2 of Chapter 1. Here, to find an optimal policy, you first need to assess the probability $P(D)$ that an individual has the disease in question. The client is encouraged to think of $P(D)$ as the proportion of patients with the disease in what he regards to be a 'typical' population of patients. A clinician may well feel quite confident in guessing this proportion even when he has not previously treated the disease in question. He will often be able to relate the assessed proportion to his experiences when he previously examined populations of patients with similar diseases.

Although the relative frequency approach can be very useful it has several drawbacks which make it inappropriate to use in many problems. It is often very difficult for the client to think of a simple and sensible population in which to embed the event of interest. In fact, if the event is a 'one-off', without hideous mental contortions it is almost impossible for a client to specify such a proportion. And even if the event of interest naturally nests itself inside a repeated experiment as in the above example the 'typical population of patients' may be very difficult to conceptualize.

A second and more fundamental criticism of this approach is that it does not specify the criterion under which the doctor's 'best guess' is the proportion he chooses. To be precise about what exactly is meant by a 'best guess' entails the introduction of lotteries of one sort or another. But the main advantage of the relative frequency approach is that it avoids you asking a doctor to make tasteless hypothetical bets about such things as the probability of a patient's survival. Anyway, it could be argued that if we are introducing lotteries we may as well use the method of assessing probabilities given in the next section which uses lotteries directly.

Despite these criticisms, clients who are used to communicating results through proportions of items, incidents or individuals will feel much more comfortable stating probabilities in terms of hypothetical relative frequencies than by relating probabilities to decisions which maximize their financial gain. Because of this you will find you obtain your best results with such clients by using relative frequency subjective probabilities.

For a further discussion of the relative frequency approach see Phillips (1982). In the clinical field one is often in the position of having large amounts of data to help assess these probability assessments. Spiegelhalter and Knill-Jones (1984) give a review of this area and present their own methodology based on 'weights of evidence' which are the logarithms of ratios of conditional probabilities (see also Good, 1950, 1983).

## 4.3 ASSIGNING A PROBABILITY TO A ONE-OFF EVENT

Let us return to considering the oil exploration example of Chapter 2. To proceed with that analysis we assumed that the probabilities of oil existing in each of the two oilfields were numbers that were given. Of course in reality these numbers would not be given to your client. In fact it is very likely that your client has no data given as a relevant frequency on which to base his probability assessment. Nor is it particularly easy to think of any one oilfield as a member of a very large population. Only a limited number of fields physically exist and these vary greatly in type.

So you are confronted with the following question: given that it is sensible to assign a probability to a 'one-off' event in a decision problem, how should you do it?

To make such an assignment you can insist that the client's subjective probabilities 'cohere' with his stated preferences over lotteries with known probabilities. To ensure that you can obtain such subjective 'probabilities' you need to do two things. First you must extend the class of options you offer your client to include a set of hypothetical betting schemes to which he can readily assign an expected utility. Second you demand that your client satisfies Rules 1–4 of Chapter 3 on this extended set of options. Using the example of Chapter 2 we shall illustrate how following such a procedure assigns a unique 'probability' to any event.

*Example 4.1* (Continuation of Example 2.1)

For simplicity and without loss of generality, assume that your client's utility is linear in pay-off so that the originally discussed EMV algorithm is appropriate, utilities being equal to pay-offs.

You start by considering the option (decision $d_4$) of drilling a well without investigating the field beforehand. Oil will be present (event $B$) or not present (event $\bar{B}$) and you can easily calculate $u(d_4|B)$ and $u(d_4|\bar{B})$ – these are just the terminal pay-offs 164 and $-31$ respectively. Now suppose your client were told that the probability of oil being found in $B$ was $\alpha$. Call this hypothetical betting scheme $b(\alpha)$. The expected utility for $b(\alpha)$ is by definition

$$\bar{u}(b(\alpha)) = \alpha 164 + (1-\alpha)(-31) = 133\alpha - 31$$

which is well defined for each value of $\alpha$ between 0 and 1.

Suppose now that your client was prepared to specify whether or not he would prefer $d_4$ to $b(\alpha)$, for each value of $\alpha$, $0 \leqslant \alpha \leqslant 1$. Clearly

$$b(0) \overset{*}{\lesssim} d_4 \overset{*}{\lesssim} b(1)$$

so if he follows the four utility rules on the set of bets including the hypothetical ones like $b(\alpha)$, then by equation (3.1) there is a *unique* value $\alpha(B)$ of $\alpha$ such that $d_4 \overset{*}{=} b(\alpha(B))$ (by Rule 4).

Under the utility rules if he were to assign a probability to the event $B$ it would have to be $\alpha(B)$. If it were anything else then bets with the same distribution would not be equally preferred and this would violate Rule 1. His 'subjective probability' of oil being present in field $B$, if it can be defined, must be equal to $\alpha(B)$.

The general procedure for finding the probability of any event in a decision problem is exactly analogous to the one given above. To assign a probability to an event $E$ whose occurrence influences your consequent pay-off distribution you compare a simple real decision rule $d$, with hypothetical bets $b(\alpha)$ having *known* probabilities $\alpha$ and $1-\alpha$ of $E$ and $\bar{E}$ respectively occurring for all

$0 < \alpha < 1$. The subjective probability $P(E)$ of $E$ is then defined as the value $\alpha(E)$ for which $b(\alpha)$ is equally preferred to $d$.

Conditional probabilities can be elicited using a similar method to the one given above for marginal probabilities. Returning to the drilling example of Chapter 2, suppose your client needs to give a subjective probability of $P(a|\bar{A})$, the probability that an investigation says oil is present in a field when it is not. For the sake of simplicity assume that it will be resolved quickly whether or not a field will contain oil and as before, and without loss of generality, that your client's utility on money is linear. To elicit $P(a|\bar{A})$ you ask about his preferences over the following two types of wager.

For the first wager $b_1$ you ask your client to imagine that he is offered a warranty on the test equipment he uses in his investigation. If the investigation advises drilling and oil is found not to be present he will be paid \$1m. If the investigation advises drilling and oil is not present he will be paid nothing and if oil is found to be present the warranty will not operate. This type of wager is called a *called-off bet* or *censored experiment* – being 'called off' if an event (in this case $A$) occurs.

For the second type of wager $b_2(p)$ the client will simply receive \$$p$m, $0 \leqslant p \leqslant 1$, if oil is not found in the field. You ask your client the maximum value of $p = p^*$ for which

$$b_2(p^*) \overset{*}{\lessgtr} b_1$$

Clearly as an EMV decision-maker, if your client knew that $P(a|\bar{A}) = \alpha$, then *given $\bar{A}$ occurred*, $b_1$ would be worth $\alpha = \alpha 1 + (1-\alpha)0$ to him. Being given \$$\alpha$ if $\bar{A}$ occurs is by definition $b(\alpha)$, so $p^* = \alpha$. Using the same arguments to the ones used for eliciting the probability of oil, we can now deduce that if the client is indifferent between $b_1$ and $b_2(p^*)$, then, if his subjective probability $P(a|\bar{A})$ can be defined at all, it must be equal to $p^*$.

Of course, by calling these artefacts of a client's preferences 'probabilities' carries implications of its own. In particular, you assume that over any event space, the probability measure $P$ will satisfy the usual axioms of probability. Whether it is reasonable to conclude this is the subject of the next section.

## 4.4 THE SUBJECTIVE PROBABILITY ASSUMPTION AND SOME CRITICISMS

However you choose to elicit subjective probabilities, implicit in their use is the following assumption.

*The subjective probability assumption*

In the problem facing your client he can unambiguously express his beliefs about the rewards arising from each decision he might take through a probability distribution of rewards labelled by that decision. If he states other subjective probabilities to help him calculate these distributions of rewards, then these probabilities can be manipulated using the usual laws of probability.

Now this assumption is far reaching and has been criticized as being too strong. It forms one of the cornerstones of Bayesian decision analysis, however. Below I give two examples illustrating the types of criticisms that have been made.

*Example 4.2*

Two tennis players, Smith and Brown, will play against each other in a tournament which is guaranteed to take place. Your client has no information about the two players except their names. He is offered two wagers where he wins \$10 if he picks the winner and \$0 otherwise. He tells you he would pay a maximum amount of $\$r_1$ to back Smith to win and, because of the symmetry of his information, a maximum amount of $\$r_1$ to back Brown to win.

Now he is offered another wager in which he wins \$10 if and only if he predicts the outcome of the toss of a fair coin. He states that he would be prepared to pay a maximum amount of $\$r_2$ either for the option of calling heads or the option of calling tails.

Provided that he employs the subjective probability assumption, by symmetry he must assign a probability $\frac{1}{2}$ to either player winning. It follows from utility Rule 1 that if his reward can be measured by his monetary gain then he must state $r_1 = r_2$. It has been observed that many people 'because they are more sure of the odds' will choose $r_2 > r_1$, thus either violating the subjective probability assumption or utility Rule 1. Notice here that if we chose to measure his subjective probability using the method as described in Section 4.3 then that subjective probability would be less than $\frac{1}{2}$ both for Smith winning and for Brown winning, yet we know that one of these players must win.

One counterargument that is often used to explain this phenomenon is that in wagers like the one mentioned above the client's reward space may not only involve his monetary gain but also his loss of face. He might be expected to be able to predict the winner in the tennis game and when he does not he may appear foolish. On the other hand, in the coin toss any failure will be attributed to his bad luck. Another counterargument to the tennis example, often used to try to convince a client that he should adjust his preferences so that $r_1$ equals $r_2$, is given below.

You offer the client above who sets $r_1 < r_2$ two new wagers. In the first, $b_1$, the client will be told tomorrow the result of the tennis match. If Smith won he will be given the option of betting on a head occurring on the toss of a fair coin, receiving \$10 if it does. On the other hand, if Brown won then he will be given the option of betting on a tail occurring on the toss of a fair coin, receiving \$10 if it does. Since a wager on a head or a tail are equally preferable to him and worth $\$r_2$ for a \$10 win, there is a very strong argument for concluding that wager $b_1$ is worth $\$r_2$ to him.

In the second wager, $b_2$, a coin is tossed before the result of the match is known to the client. If a head results then the client will win \$10 if Smith wins and if a tail results will win \$10 if Brown wins. Since he is indifferent between the two bets on the tennis match, each being worth $\$r_1$, surely $b_2$ should be worth $\$r_1$ to the client.

But now we have come to an apparent contradiction. For both wager $b_1$ and wager $b_2$ will win the client \$10 if and only if a head and a Smith win or a tail and a Brown win occurs. So surely they should be worth the same amount to the client, for otherwise the client's beliefs would depend on the order he will receive information.

Suffice to say that not all theorists have been convinced by the counterargument above and some have attempted to accommodate a preference for bets with 'known probabilities' into a formal framework (see e.g. Schmeidler, 1984). Even if you find these counterexamples convincing you, as a decision analyst you are left with the practical problem that sometimes your elicited one-off probabilities will be too conservative.

A second argument commonly used against the subjective probability assumption is that it is not always possible to represent ignorance within it.

*Example 4.3*

A client believes that a random variable $X$ has a Poisson distribution with mean $\theta > 0$. He believes $\theta$ is equally likely to lie in any two intervals $I_1$ and $I_2$ which have the same length and are contained on the positive real line. Unfortunately no probability distribution exists that represents this belief.

To illustrate, let $I_i = (i-1, i] \qquad i = 1, 2, 3, \ldots$

Now we have $P(I_i) = P(I_j) \qquad i, j = 1, 2, 3, \ldots$

However, since these intervals are disjoint, by the laws of probability we must have simultaneously that

$$\sum_{i=1}^{\infty} P(I_i) = 1.$$

Clearly these two statements cannot both hold at once.

In a decision analytic setting where $X$ is a future observable in a well-specified problem it would be unusual for a client to be totally ignorant about $\theta$ in this way. On the other hand, in a statistical problem where a client may well not wish to use any information about $\theta$ even if he has it, Example 4.3 poses a more worrying problem. One partial solution is to impose only 'finite additivity' on subjective probability distributions. Unfortunately such a treatment is beyond the scope of this book (see De Finetti, 1974).

Because the subjective probability assumption is contentious, many authors have attempted to justify it by various types of axiomatic systems which discuss the appropriateness of each of the axioms of probability in turn. Each of these axiomatic systems is designed with its own method of elicitation in mind. Full discussions are given in, for example, De Finetti (1974), Savage (1954), DeGroot (1970) and French (1986). A good text for comparing the different axiomatic systems is Fine (1973). Two accessible discussions on given types of subjective probability are given in Lindley (1985b) and Raiffa (1968). Exercise 4.1 at the end of this chapter illustrates one of the types of argument used.

In practice the probabilities you elicit will invariably *not* satisfy the probability axioms if you happen to ask enough questions, though the client is usually prepared to adjust them when you point this out and treat the inconsistencies as measurement errors. If he is not prepared to do this it may be either that it is not possible to express his beliefs using the subjective probability assumption or, more usually, that the whole structure of the model you are imposing on your client is wrong (see Chapter 5). Automatic ways of adjusting subjective probabilities so that they *cohere* – i.e. satisfy the laws of probability, are given in Lindley, Tversky and Brown (1979).

Personally I have found the subjective probability assumption is a very useful practical tool. On theoretical grounds it seems to me to be no more or less questionable than the assignment of utilities – a procedure which has acquired a large degree of academic respectability. So to reject the idea that a client's opinions can be expressed probabilistically should also encourage the rejection of the conventional utility axioms. Wolfenson and Fine (1981) do exactly this by constructing new axiomatic systems based on the idea that a client cannot reasonably be expected to always be able to find a betting scheme which he considers equivalent to a given reward. As a consequence they argue that Rules 1–4 given in the last chapter should be broken. In contrast, Goldstein (1981) defines an inferential structure based on the specification of a few expectations rather than the specification of whole probability distributions. This structure is not quite rich enough, however, to solve general decision problems. An alternative approach using 'beliefs functions' is presented in Shafer (1982).

All these alternative approaches have their own drawbacks. For example, none can guarantee imposing a total ordering on a decision space. The approach given here is currently the most popular, and appears to work quite well in practice.

## 4.5 HOW TO ASSESS SUBJECTIVE PROBABILITIES AS ACCURATELY AS POSSIBLE

Let us assume that you are happy about assigning subjective probabilities to your client's beliefs. In normal circumstances, because of lack of resources, you will only be able to ask him to give subjective probabilities for a limited number of events. You will then implicitly or explicitly deduce your client's probabilities of other events of interest, using the usual rules of probability.

Now it has already been stated in this chapter that it is usually the case that if you were to ask a client for all probability statements associated with an event space, then the probabilities would not cohere. He would later want to adjust some of his statements so that they did. Therefore if you are going to ask him for only a relatively small number of probability statements you should ask for those which you believe would be least likely to be adjusted in the light of seeing other probabilities you have not asked for. The five principles below, although no panacea, should help you obtain probability assessments from your client which are fairly reliable.

*Principle 1*

Deduce probabilities of 'one-off' events using betting schemes which are as close as possible to your client's problem. If a client cannot understand why you are asking for a preference he will not expend much energy answering it. On the other hand, if he can see that the preference is important he will concentrate harder and give a more realistic answer.

*Principle 2*

If your time resources allow you, try to break down an event into component events (this is called the *principle of credence decomposition*).

In Example 1.2 the client's optimal policy is dependent on the proportion $P$ of patients he had seen who were expected to have the disease. Because of lack of data this probability may well need to be a subjective probability. Suppose, for the sake of argument, that the disease is most prevalent in pregnant women and patients over the age of 60. Let $A$ denote the event that any one patient is pregnant, $B$ denote the event that a patient is over the age of 60, and $C$ denote the event that a patient is in neither of these classes. The probability $P(D)$ that an individual has the disease can be broken down by 'extending the conversation'. Thus:

$$P(D) = P(D|A)P(A) + P(D|B)P(B) + P(D|C)P(C)$$

The client may well be able to assess accurately the probabilities that a patient lies in one of the classes $A, B$ or $C$ and the probabilities of the disease given that the patient lies in these classes in his designated sample. He may have great difficulty specifying $P(D)$ directly because he will need to try and 'extend the conversation' in his head – quite a difficult manipulation!

Note that it is probably most efficient to use the 'relative frequency approach' to assess probabilities in this problem. For example, to assess $P(A)$ he needs to predict the proportion of pregnant women he expects to see in his class of all patients.

Another decomposition which is often useful breaks down an event $E$ into various conditional probabilities. Thus

$$P(E) = P\left(\bigcap_{i=1}^{n} E_i\right)$$

$$= \prod_{k=2}^{n} P\left(E_k \middle| \bigcap_{i=1}^{k-1} E_i\right) P(E_1) \qquad (4.1)$$

For example, suppose you wanted to calculate the probability that a particular nuclear reactor will blow up in the next five years (event $E$). This event would only occur if:

(a) the conditions arose which could cause a blow-up (event $E_1$);
(b) a number $n-1$ of safety devices all simultaneously failed to detect and defuse the potential disaster (events $E_2, E_3, \ldots, E_n$).

Rather than attempting to calculate $P(E)$ directly it is usually more accurate to assess $P(E)$ using equation (4.1). So you first evaluate $P(E_1)$ and then evaluate $P(E_{k+1}|E_1, E_2, \ldots, E_k)$, the probability that the $k$th safety device fails given $E_1$ and given the first $k-1$ safety devices have all failed, $1 \leqslant k \leqslant n-1$.

A third common type of decomposition is possible when you are trying to assess the probability *distribution* of a certain random variable. Suppose that a wholesaler needs to assess the probability distribution $F_x$ of his turnover $X$ over the next month. Rather than trying to assess $F_x$ directly he decomposes the random variable $X$ into its components. Thus if he sells $n$ items the total turnover $X$ is given by

$$X = \sum_{i=1}^{n} C_i R_i \qquad (4.2)$$

where $C_i$ is the price of the $i$th item he sells and $R_i$ is the number of the $i$th item he sells, $1 \leqslant i \leqslant n$. He assesses the probability distributions of $R_i$, $1 \leqslant i \leqslant n$, and then deduces the distribution of $X$ from equation (4.2). Since he will know $C_i$, $1 \leqslant i \leqslant n$, and $R_i$ will vary dependent on his order book, it will usually be more accurate to calculate the distribution of $X$ using this decomposition rather than directly.

*Principle 3*

Whenever possible do not try to directly ascertain probabilities of events which you expect your client to state as close to 0 or 1. It is usually possible to avoid the assessment of very large or very small probabilities by using a suitable credence decomposition.

The nuclear reactor example discussed above achieves this by the conditional decomposition of equation (4.1). It has been widely observed that people are very poor at assessing probabilities close to zero or one.

*Principle 4*

If $A$ causes $B$ (in the sense that $A$ 'partially causes' rather than 'logically causes' $B$) then it is often better to assess the joint probability from the formula

$$P(A \cap B) = P(B|A)P(A)$$

rather than the formula

$$P(A \cap B) = P(A|B)P(B)$$

*Example 4.4*

Suppose a client owns a warehouse in which he has installed a sprinkler system to douse the flames of a possible fire. Let $A$ be the event that a fire occurs in the next five years and $B$ the event that the sprinkler system is unsuccessful in putting out the fire. He is interested in $P(A \cap B)$, the probability that his goods will be destroyed by fire. Ideally the fire should *cause* the sprinkler to work. So to assess $P(A \cap B)$ first assess your client's probability $P(A)$ that a fire occurs. Then assess the probability $P(B|A)$ that the sprinkler system fails to work given that a fire occurs. Notice how difficult it is to think about $P(A|B)$, the probability of fire given the sprinkler system has failed!

## 4.6 CAUSAL DIAGRAMS

So far we have only considered decompositions of an event or random variable into simple component events or random variables. However, quite often the causal links underpinning the occurrence or otherwise of an event are complex. This is particularly true when an event of interest depends on the interactions of other people or institutions over whom you have little or no control. To follow the causal Principle 4 you will then need a more subtle decomposition of your event. Providing a causal diagram can help you achieve an appropriate decomposition.

*Example 4.5*

The Home Office needs to decide whether to buy a stock of new riot equipment this year for a particular high-security prison S. They therefore need to assess the probability that a riot will occur (event $E$) next year in S. The occurrence of $E$ will depend on the values of several random variables. Some of the more important factors are listed below.

1. a measure ($X_1$) of the ability of S to detect and defuse a potential riot;
2. a measure ($X_2$) of the aggrievances felt by the inmates;
3. a measure ($X_3$) of the aggrievance felt by the prison officers;
4. a measure ($X_4$) of inmate overcrowding in S;
5. a measure ($X_5$) of the rate of change of inmate population in the next year;
6. a measure ($X_6$) of the degree of change in regime in S next year.

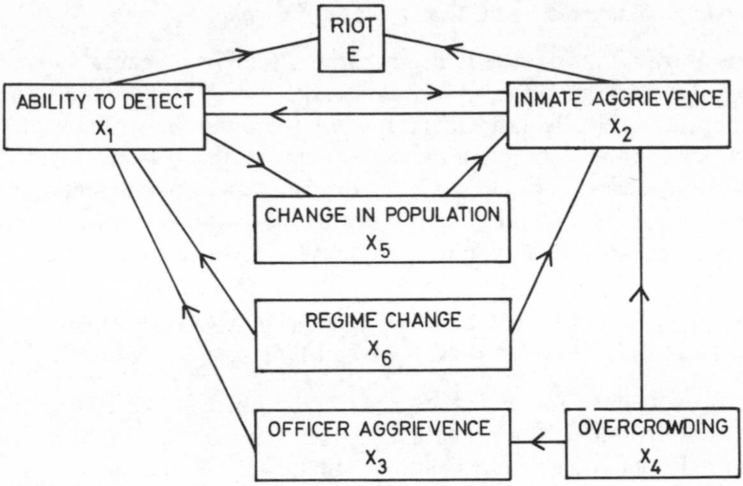


**Figure 4.1** Causal diagram for the probability of a riot

This can be represented by the causal diagram given in Fig. 4.1. An arrow from box $A$ to box $B$ signifies that $A$ causally effects $B$ to some greater or lesser extent. For example, overcrowding ($X_4$) tends to cause bad working conditions for prison wardens (influencing $X_3$) and make it more difficult for inmates to cope with the prison regime (influencing $X_2$). If you believe that both $A$ causes $B$ and $B$ causes $A$, simply draw two causal arrows, one in each direction. This is illustrated in Fig. 4.1 for the link between $X_1$ and $X_2$.

To follow Principle 4 you start by assessing the distributions of random variables which are not 'caused by' other random variables in your diagram. For the causal diagram of Fig. 4.1 you therefore start by assessing the joint distribution of $X_4$ and $X_6$. Next assess the distribution of random variables which are only influenced by $(X_4, X_6)$. The only such random variable in Fig. 4.1 is $X_3$. Now calculate the distribution of $(X_3|X_4, X_6)$ for each possible value of $X_4$ and $X_6$. Ideally you would now find the conditional distribution of those random variables 'caused' only by $(X_3, X_4, X_6)$. Unfortunately in our example there are no such random variables. However, $X_1$ is only 'caused by' one of the remaining variables, $X_2$ so to violate Principle 4 least it is probably wisest to assess the probabilities of $(X_1|X_3, X_4, X_6)$ next. Continue by assessing $(X_5|X_1, X_3, X_4, X_6)$ and then $(X_2|X_1, X_3, X_4, X_5, X_6.)$ Finally calculate $P(E|\mathbf{X})$ for all possible values of $\mathbf{X}$ where $\mathbf{X} = (X_1, X_2, \ldots, X_6)$.

Clearly there is a great deal of work required to devise a probability in this way before initiating such a task. It is therefore wise to consider whether the probability of an event needs to be known very accurately. Credence decomposition will be discussed further in Chapter 6. Ideas of necessary accuracy are considered in the next section.

*Two cautionary notes on the use of causal diagrams*

1. It is always important to remember that if there is no causal link between two random variables, it does not necessarily follow that they are independent. In the last example $X_4$ and $X_6$ may be strongly correlated with one another although there is no causal link between them.
2. Causality occurs in real time. Causal diagrams give no representation of this temporal aspect of causation. By ignoring the dynamic aspects of a problem they sometimes direct you to a poor breakdown of a problem.

## 4.7 SENSITIVITY ANALYSIS IN THE ASSIGNMENT OF SUBJECTIVE PROBABILITY

Suppose you have followed the principle outlined in the last section and have obtained subjective probabilities from your client to find his optimal decision. Either you or your client may still be worried that either the utilities or the probabilities in the problem have been measured inaccurately and that these inaccuracies may direct you to choose a decision which is very suboptimal.

As computers become more powerful and accessible it is becoming possible, even for complex decision problems, to investigate numerically how a client's optimal decision, and consequent reward distribution, are influenced by the values he assigns to his utilities and probabilities.

For example, suppose the client's expected utility conditional on his subjective probability statements $\boldsymbol{\pi}=(\pi_1,\pi_2,\ldots,\pi_n)$ is $\bar{u}(d|\boldsymbol{\pi})$ for a decision $d$ in his decision space $D$. The values of $\bar{u}(d|\boldsymbol{\pi})$ can be calculated for each value of $d \in D$ on a fine lattice of different values of $\boldsymbol{\pi}$ in a neighbourhood of the original subjective probabilities. It is then possible to see how the Bayes decision varies as $\boldsymbol{\pi}$ is perturbed from its initial value. If you find that the Bayes decision and its associated pay-off distribution are very sensitive to some component $\pi_j$ of $\boldsymbol{\pi}$, then a case can be made for trying to measure $\pi_j$ more accurately. For many problems you will find that the Bayes decision is surprisingly robust against quite large perturbations of $\boldsymbol{\pi}$. This fact is often very reassuring to the client who can see that his choice of optimal policy does not depend very heavily on the accuracy of his personal input into the model.

Clearly a similar procedure can be adopted to investigate how perturbing a utility function may effect the client's optimal policy.

## 4.8 CALIBRATION AND SUCCESS AT MAKING PROBABILITY STATEMENTS

How good are clients at giving a decision analyst accurate probability statements? Well in the first part of this chapter we discussed why the accuracy of probability assessment was very dependent on the way you formulate your questions. However, if a client is making a sequence of probability forecasts about a specific type of occurrence he is interested in, it is then possible to obtain an *empirical* measure of that client's accuracy. This is because we can then compare his forecasts with the frequency with which events actually occur.

For simplicity, we shall consider only one type of probability forecaster – the weatherman. In the United States, every day weather forecasters from different television channels give probabilities of rain in the next day at various locations. Rain is defined as at least a fixed precipitation of water occurring in a fixed length of time. Obviously yesterday's prediction can be compared with whether or not it rained today.

Let us set up some notation. Let $X_t$ denote the randc m variable which takes the value 1 if it rains tomorrow (time index $t$) and 0 if it does not. Let $q_t$ denote a weatherman's stated probability of rain, quoted today, that it will rain tomorrow. Clearly for $t = 1, 2, 3, \ldots, 0 \leqslant q_t \leqslant 1$. We can now give the following definition.

*Definition*

A weatherman is said to be *empirically well-calibrated* over $n$ time periods, if over the set of days he quotes the probability of rain as $q$, the proportion $\hat{q}$ of those days that are rainy days is $q$, this being true for all values of $q$ he quotes.

Let $n(q)$ denote the number of times a weatherman quotes $q$ and $r(q)$ denote the number of times it rained given he made the probability forecast $q$. Then empirical calibration say that $r(q)/n(q) = q$ for all quoted $q$. Table 4.1 gives

**Table 4.1** The performance of four weathermen in a hypothetical experiment to check their calibration. The number of forecasts $q$ and the number of rainy days given forecast $q$ are respectively $n(q)$ and $r(q)$.

| | | Probabilities ($q$) quoted | | | |
|---|---|---|---|---|---|
| Weathermen | $(n, r)$ | 0 | 0.5 | 1 | Well calibrated |
| $W_1$ | $n(q)$ | 20 | 60 | 20 | √ |
| | $r(q)$ | 0 | 30 | 20 | |
| $W_2$ | $n(q)$ | 50 | 0 | 50 | √ |
| | $r(q)$ | 0 | 0 | 50 | |
| $W_3$ | $n(q)$ | 40 | 10 | 50 | × |
| | $r(q)$ | 5 | 5 | 40 | |
| $W_4$ | $n(q)$ | 0 | 100 | 0 | |
| | $r(q)$ | 0 | 50 | 0 | √ |

the result of a hypothetical experiment in which four weathermen were asked to predict whether it would rain on each of 100 consecutive days. All four forecasters happen only to have quoted probabilites 0, 0.5 and 1 and it happened to rain on exactly 50 days out of the 100 days observed.

Weather forecasters $W_1, W_2, W_4$ are all empirically well calibrated since $r(q)/n(q) = q$ for $q = 0$, 0.5 and 1 for all three. Forecaster $W_3$ is not well calibrated because $r(1)/n(1) = 0.8 \neq 1$, for example.

The reason why calibration is an important concept is that it can be proved that if $q$ were the 'true' probability of rain whenever it was quoted then

$$\frac{r(q)}{n(q)} \to q \quad \text{as } n(q) \to \infty \quad \text{with probability 1}$$

So if a weatherman is accurate in his probability forecasts you know that in the long run (provided he doesn't quote too many different values of $q$ so that $n(q)$ is not small) he will be approximately empirically well calibrated.

Unfortunately not all well-calibrated weathermen give useful forecasts. Compare weatherman $W_2$ with weatherman $W_4$ of Table 4.1. $W_2$ makes the right prediction every day whereas $W_4$ *always* says there is a 50% chance of rain. $W_2$ gives very useful forecasts whereas $W_4$ forecasts are next to useless. All we can say about calibrated forecasters is that they are realistic about their own abilities to forecast.

For more details on the subject of calibration see Dawid (1982, 83, 84) or DeGroot and Fienburg (1983).

## 4.9 ARE PEOPLE WELL CALIBRATED?

We now turn to the question: 'In practice, are probability forecasters approximately well calibrated or do they state probabilities which bear little relationship to frequencies of events?' To investigate this question the proportion $\hat{q}$ of events occurring when assigned a probability $q$ by the client is plotted against $q$ where $q$ has been rounded to the first decimal place (say). The resulting graph is called his *calibrated curve*. If he were approximately empirically well-calibrated then $\hat{q}$ would lie approximately on the straight line $\hat{q} = q$. On the other hand, if $\hat{q}$ appears approximately constant in $q$ this would indicate that the client's probability statements were misleading and uninformative.

It has been discovered that a client's abilities to forecast probabilities depend very heavily on both the type of client and the events he is being asked to forecast. On the positive side, people who are used to dealing with repeated uncertain events tend to produce very reliable probability forecasts. In particular, bookmakers and weathermen are good probability forecasters. Figure 4.2 shows two observed calibration curves for weathermen making short-term wheather forecasts. Notice that both lines are very close to $\hat{q} = q$.

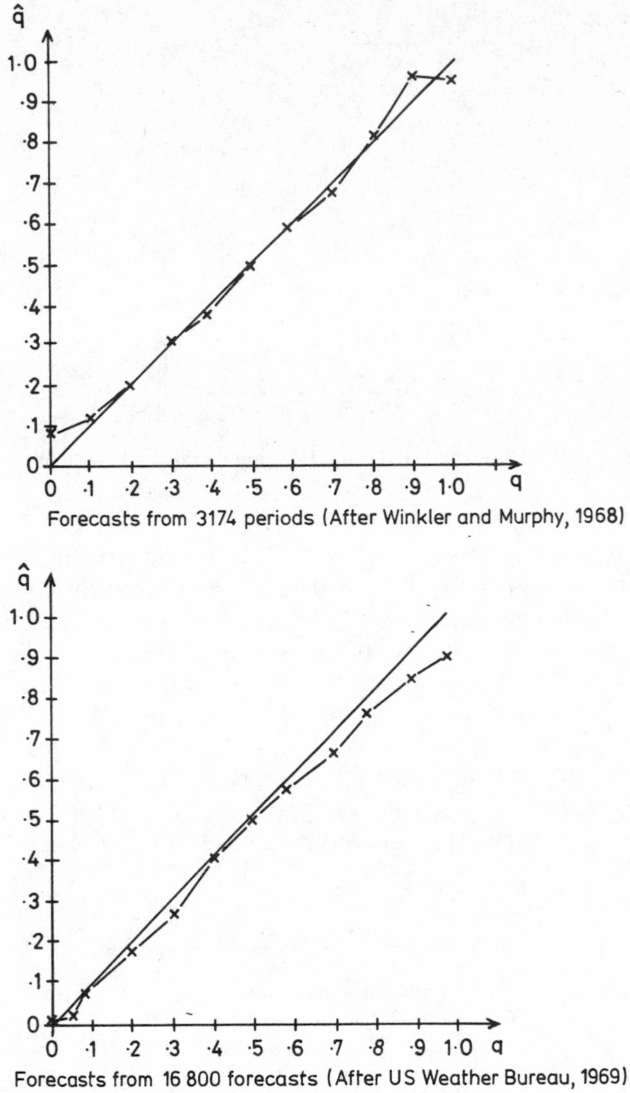


**Figure 4.2** Two typical calibration curves for weathermen's forecasts of rain

Major deviations seem to have occurred only when the probability of rain is perceived to be very small or very large. Apparently since these experiments have been communicated to weathermen they now perform much better even for these large and small probabilities.

On the negative side, people with no substantive knowledge of the problems they are being asked about tend to be very poorly calibrated. This was

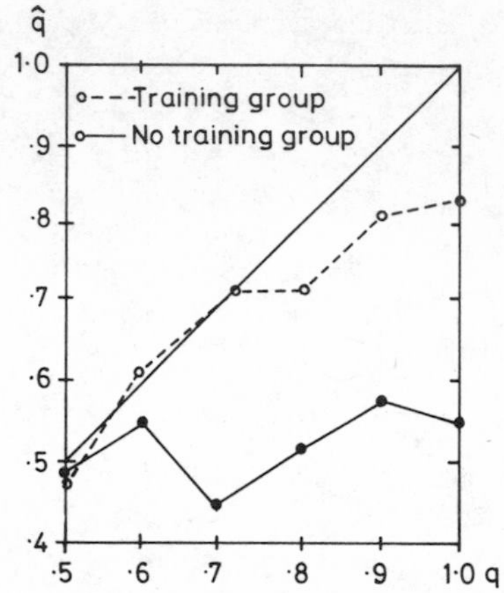


**Figure 4.3** Probability forecasting of handwriting (after Lichtenstein and Fischoff, 1976)

illustrated when subjects were asked to identify the country of origin of people's handwriting (see Fig. 4.3). On the other hand, the subjects' calibration improved dramatically when they were given a little training on how to distinguish between the two groups in the test.

Problems also occur when clients are not culturally used to the numerical quantifications of uncertainty. Figure 4.4 records the failure of Chinese to give informative probability forecasts.

The typical businessman's ability to forecast probabilities has been found to lie somewhere between these two extremes. Certainly knowledgeable businessmen will usually provide informative if not completely calibrated

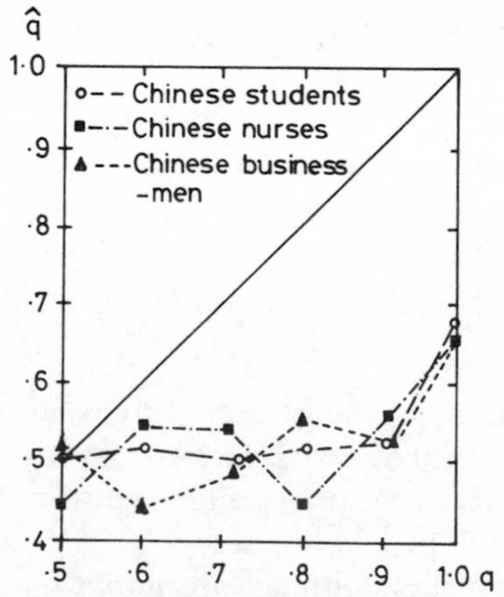


**Figure 4.4** Calibration of Chinese (after Phillips and Wright, 1977)

forecasts. Furthermore, their performance tends to improve dramatically when their calibration curves are fed back to them. In my opinion this gives qualified encouragement for the Bayesian approach of persuading a client to express his beliefs probabilistically. For a much more detailed summary of some of these results see Lichtenstein *et al.* (1977, 1982) and Phillips and Wright (1977).

## 4.10 SCORING PROBABILITY FORECASTERS

I have indicated that calibrated forecasters do not necessarily produce useful forecasts. However, you can measure forecasters' abilities in the following way. Suppose you reward a forecaster for producing good forecasts. His stated probability $q$ can then be regarded as a decision. Explicitly, suppose that every day you reward a forecaster an amount $u(x, q)$ when he has made a forecast $q$, and $x = 1$ if it rains that day and $x = 0$ otherwise. The function $u(x, q)$ is called a *scoring rule*. If you were to make the heroic assumption that the weatherman wants to use the EMV algorithm he should choose a forecast $q$ which maximizes

$$\bar{u}(q \mid p) = pu(1, q) + (1 - p)u(0, q)$$

where $p$ is his subjective probability that rain will occur tomorrow.

*Definition*

Any pay-off function $u(x, q)$ for which $\bar{u}(q \mid p)$ is (uniquely) maximized when $q = p$ for any value of $0 \leqslant p \leqslant 1$, is called a *(strictly) proper scoring rule* [(s)psr].

A psr encourages honest forecasting in that if a weatherman is acting optimally, in the sense given above, he should quote his subjective probability of an event.

*Example 4.6*

**Brier score** $\quad u(x, q) = 1 - (x - q)^2$

$$\bar{u}(q \mid p) = 1 - p(1 - q)^2 - (1 - p)q^2 = 3/4 + (p - \tfrac{1}{2})^2 - (q - p)^2$$

For a fixed value of $p$, $u$ is clearly uniquely maximized when $q = p$. Therefore the Brier score is an spsr. Note that your client expects his worst pay-off, 3/4, when the true probability $p = \frac{1}{2}$.

**Logarithmic score** $\quad u(x, q) = \begin{cases} \log q & \text{if } x = 1 \\ \log(1 - q) & \text{if } x = 0. \end{cases}$

$$\bar{u}(q \mid p) = p \log q + (1 - p) \log(1 - q)$$

Differentiating this expression with respect to $q$ and equating to zero gives us that $\bar{u}(q|p)$ is uniquely minimized when $q=p$, $0 \leqslant p \leqslant 1$. Hence the logarithmic scoring rule is an spsr.

**Absolute score** $$u(x, q) = 1 - |x - q|$$

This is not a proper scoring rule. It can be shown that $\bar{u}(q|p)$ is maximized at $q^*(p)$ where

$$q^*(p) = \begin{cases} 1 & \text{if } p > \frac{1}{2} \\ 0 & \text{if } p < \frac{1}{2} \end{cases}$$

The weathermen in Table 4.1 can now be compared by calculating their aggregate winnings (called the *Brier score aggregate*):

$$S_n = \sum_{i=1}^{n} u(x_i, q_i)$$

where $n = 100$. Under the Brier score it is easily checked that the aggregate scores $S_n$ of $W_1, W_2, W_3, W_4$ are respectively 85, 100, 82.5 and 75. Under the Brier score, therefore, the order of merit of our forecasters is $W_2, W_1, W_3, W_4$.

## 4.11 A LINK BETWEEN THE BRIER SCORE AND EMPIRICAL CALIBRATION

The following theorem shows that we can often improve the score of a weatherman by recalibrating him.

*Theorem 4.1*

A forecaster $P$ quotes one of $m$ probabilities $q_j$, $n(q_j)$ times, $1 \leqslant j \leqslant m$. Let $\sum_{j=1}^{m} n(q_j) = n$, the total number of days of the experiment. $P$'s Brier score aggregate $S_n$ would be at least as large if his forecasts, $q_j$, were replaced by their sample proportion

$$\frac{r(q_j)}{n(q_j)} = \hat{q}_j$$

(say) where $r(q_j)$ is the number of times it rains when $P$'s forecast is $q_j$.

**Proof**

$$\begin{aligned} S_n &= n - \sum_{i=1}^{n} (x_i - q_i)^2 \\ &= n - \sum_{j=1}^{m} \sum_{k=1}^{n_j} (x_{k,j} - q_j)^2 \end{aligned}$$

where $n_j = n(q_j)$ and $x_{k,j}$ is the result of the $k$th day when $P$ assigns a probability $q_j$.

Now

$$\sum_{k=1}^{n_j} (x_{k,j} - q_j)^2 = \sum_{k=1}^{n_j} [(x_{k,j} - \hat{q}_j) + (\hat{q}_j - q_i)]^2$$

$$= \sum_{k=1}^{n_j} (x_{k,j} - \hat{q}_j)^2 + n_j(\hat{q}_j - q_j)^2$$

Since

$$\hat{q}_j = \frac{r(q_j)}{n(q_j)} = \sum_{k=1}^{n_j} \frac{x_{k,j}}{n_j}$$

by the definition of $x_{k,j}$ so that the cross-term

$$2(q_j - q_j) \sum_{k=1}^{n_j} (x_{k,j} - \hat{q}_j) = 0.$$

Thus

$$S_n = n - \sum_{j=1}^{m} \left[ \sum_{k=1}^{n_j} (x_{k,j} - \hat{q}_j)^2 + n_j(\hat{q}_j - q_j)^2 \right]$$

$$= S'_n + \sum_{j=1}^{m} n_j(\hat{q}_j - q_j)^2 \qquad (4.3)$$

where

$$S'_n = n - \sum_{i=1}^{m} \sum_{k=1}^{n_j} (x_{k,j} - \hat{q}_j)^2$$

$$= n - \sum_{j=1}^{n} (x_i - \hat{q}_i)^2$$

is the Brier score for $P$ if he used the sample proportions $r(q_j)/n(q_j) = \hat{q}_j$ as his probability forecasts. Since the second term in (4.3) is non-negative,

$$S_n \geqslant S'_n$$

and our theorem is proved.

Thus the Brier score aggregate of an empirically uncalibrated forecaster can always be improved retrospectively by choosing to believe a probability $\hat{q}$ whenever he quotes a probability $q$. Of course, using this type of recalibration for predicting future rainy days may not give you good results in Brier score aggregates in future days. Your forecaster may just have been unlucky early on and actually be quoting sensible probabilities. There is at least one situation when it might be sensible to recalibrate a forecaster, however. If you believe that your weatherman's own utility on his quoted probability is not linear in pay-off and does not correspond to a proper scoring rule then a recalibration of the form given above may well give you better scores in the future.

After recalibrating weatherman $W_3$ so that $q = 0$ is replaced by $q = 0.125$ and $q = 1$ by $q = 0.8$, we find his new Brier score is 85.125 and he becomes the second best forecaster of our four.

## EXERCISES

4.1 A common way of illustrating why subjective 'lottery' probabilities should satisfy the usual probability axioms is to introduce a *Dutch book* betting scheme as illustrated below. Suppose first that a client's utility is linear in monetary pay-off and the lottery probabilities of an event $E$ and its complement $\bar{E}$ are given by $P(E) = \alpha$ and $P(\bar{E}) = \beta$.

(a) Let $\alpha + \beta > 1$. Show that the client should be prepared to take on the betting scheme which assigns a gain $l\alpha^{-1}(1 + \alpha)$ with probability $\alpha$ and a loss $-l((1 - \alpha)^{-1}(1 - \varepsilon))$ with probability $(1 - \alpha)$ where $\varepsilon > 0$ is chosen sufficiently small and $l > 0$. Show that he should also accept the betting scheme which assigns a loss $l((1 - \beta)^{-1}(1 - \varepsilon))$ with probability $1 - \beta$ and a gain of $l\beta^{-1}(1 + \varepsilon)$ with probability $\beta$, where $l$ and $\alpha$ satisfy the conditions above. Show, however, that if the client accepts these two bets simultaneously he will suffer a certain loss.

(b) Construct a betting scheme which will be accepted by a client but will entail him surely losing when $\alpha + \beta < 1$.

(c) Show that if the client has a utility function which is differentiable at zero, his utility on pay-off zero, and if $\alpha + \beta > 1$ or $\alpha + \beta < 1$ then there will be two lotteries which he would volunteer to accept but if accepted *simultaneously* would be less preferable to status quo. (This type of argument can be used to show that $P(E) + P(\bar{E}) = 1$ for a 'rational' client's stated probabilities $P$ on *any* event $E$ and its complement $\bar{E}$. Similar 'Dutch book' lotteries can be constructed to show that a sure loss results if the other axioms of probability are violated.)

4.2 Show that, in the notation of Section 4.10, the Brier score aggregate $S_n$ of a forecaster $P$ can be written in the form

$$S_n = n(1 - \bar{q}(1 - \bar{q})) + \sum_{j=1}^{m} n(q_j)(\hat{q} - \bar{q})^2 - \sum_{j=1}^{m} n(q_j)(q_j - \hat{q})^2$$

where $\bar{q} = n^{-1}\sum_{j=1}^{m} n(q_j)\hat{q}_j$ is the proportion of rainy days in the whole forecasted sequence. Hence show that a calibrated forecaster $P_1$ has a greater Brier aggregate score than a calibrated forecaster $P_2$ on the same set of $n$ days if and only if $P_1$ has greater 'resolution' than $P_2$ where, under the notation above, the resolution of a forecaster is defined by

$$R_n = n^{-1} \sum_{j=1}^{m} n(q_j)(\hat{q}_j - \bar{q})^2$$

(Generalizations of these results can be found in Murphy, 1973.)

4.3 The table below gives the frequencies of probabilities of rain on the next day quoted by three weathermen $P_1, P_2$, and $P_3$, and the actual frequencies of rainy days given their forecasts over a period of 1000 days. Which of these forecasters are empirically well calibrated?

| Quoted probability of rain tomorrow | 0.0 | 0.1 | 0.2 | 0.3 | 0.4 | 0.5 | 0.6 | 0.7 | 0.8 | 0.9 | 1.0 |
|---|---|---|---|---|---|---|---|---|---|---|---|
| $P_1$'s frequency of forecasts | 0 | 0 | 0 | 0 | 80 | 580 | 340 | 0 | 0 | 0 | 0 |
| Number of actual rainy days given these forecasts | 0 | 0 | 0 | 0 | 32 | 290 | 204 | 0 | 0 | 0 | 0 |
| $P_2$'s frequency of forecasts | 420 | 0 | 0 | 0 | 0 | 60 | 0 | 0 | 0 | 0 | 520 |
| Number of actual rainy days given these forecasts | 20 | 0 | 0 | 0 | 0 | 30 | 0 | 0 | 0 | 0 | 476 |
| $P_3$'s frequency of forecasts | 20 | 80 | 125 | 120 | 80 | 100 | 60 | 120 | 155 | 90 | 50 |
| Number of actual rainy days given these forecasts | 0 | 8 | 25 | 36 | 32 | 50 | 36 | 84 | 124 | 81 | 50 |

Calculate the Brier score of these three weathermen and show that $P_2$ performs best under this criterion. By how much can $P_2$'s score be improved by recalibrating his quoted probabilities?

# 5

# Influence diagrams, group decisions, and some practical problems in decision analysis

## 5.1 INTRODUCTION: SOME PRACTICAL DIFFICULTIES IN ANALYSING DECISIONS

So far we have discussed only decision problems which are well defined and are formally possible to analyse in a rigorous and well-structured way. Unfortunately in practice many problems are not well defined.

There are several reasons for this. Sometimes a decision analyst will discover that the only reason she is being employed is to be a scapegoat for any bad decisions that the client might make. The client is not really interested in solving his problem but only in ensuring that he has someone to blame when things go wrong. Because he is only tangentially interested in solving his problem (perhaps he knows no solution can exist!) any information he provides has to be treated with scepticism. If you are forced to work for such a client, then the only suggestion I can make is that you hedge your advice in such a way to put the onus of his decision back onto him.

*Type 1 Problem What is the structure of the problem?*

Let us suppose that you are trying to solve a problem that is not of the type mentioned above. It is very common for a client to approach you simply because he is overwhelmed by the complexity of his problem. Then your primary task is to help him unravel his beliefs about the relationship between the various factors in his problem. You must also help him deduce what facets of the problem are important and which might be safely ignored in any formal analysis. He can then report back to his superiors with a well-articulated problem and they can make a decision in the light of his advice. In this situation any formal analysis you are able to perform using assessed probabilities and utilities will just help him to be more specific in his advice and may not be necessary.

Alternatively, a client may be quite confident that he understands the structure of his problem. However, a decision analyst often discovers at the *end* of an analysis that her client has forgotten to include in the description of his problem variables which in the end play the biggest role in determining the outcomes he is interested in.

Both these difficulties arise because the initial structure of the problem, or *model*, has not been thought about deeply enough. Although the decision analyst can offer no foolproof method to avoid using the wrong model, she does have in her armoury various pictorial representations of a stated problem which can be examined and discussed before any formal analysis takes place.

We have already discussed one schematic representation of a decision problem in Chapter 2, namely the decision tree. Although useful, there are many situations in which the decision tree is very unwieldy. In Section 5.2 we will study a more compact pictorial description of the stated structure of a problem which practitioners often find easy to understand. These are called *influence diagrams*. I recommend that you make it automatic practice to first represent any decision problem by its decision tree or influence diagram and discuss this with the client before embarking on any more technical analysis. This will help to minimize the possibility that you spend a large amount of time analysing the wrong problem.

*Type 2 Problem Mis-specified probabilities and utilities*

Let us suppose that you are now satisfied that you are attempting to solve the right problem. To find an optimal policy you first need to measure your client's utilities and probabilities. Although we discussed in Section 4.1 how best to make measurements of probabilities, you cannot be certain that your measurements accurately reflect your client's beliefs. The same is true of your measurement of his utility.

These difficulties are usually surmountable. Firstly, it may not matter that a client's probability and utility statements are unreliable. A 'sensitivity analysis' (see Section 4.7) may show that your decision, which was optimal under the stated probabilities and utilities, remains optimal when these probabilities are varied substantially from their stated values. Even when this is not the case you should at least be able to identify a small number of decision rules which are optimal for any plausible pair of probabilities and utility function. Furthermore, you should be able to discover which probabilities and utilities need to be known precisely before an optimal decision rule can be identified. More effort can then be expended to measure these quantities more accurately.

In practice, even if a wrong decision is chosen because of mis-specified probabilities and utilities, it is often the case that the associated reward distribution is very close to that of the optimal rule so that not too much is lost.

There is often a second way of surmounting problems of mis-specification of probabilities and utilities. It may be possible to demonstrate *analytically* that only one of a small class of decision rules could possibly be optimal regardless of the client's utility function. You will encounter an example of such an analysis in Sections 7.6 and 7.7. This set of policies can be given to the client as a partial solution to his problem.

Thirdly, a client may have made a long sequence of probability statements about events which have either occurred or not in the past. It is then possible to 'calibrate' his probability statements against observed frequencies (see Section 4.8).

*Type 3 Problem The aggregation of beliefs and preferences*

A further practical problem confronts the decision analyst when she is not employed by a single client but a committee of clients. Each member of this committee may have his or her own beliefs about the problem in question. They may also disagree about which utility function should be used. The decision analyst's task is then to find a policy which is agreeable to all members of the group.

Mutually agreeable policies are often not 'optimal' policies in the sense that we have used so far. More specifically, in general, decision rules which are chosen to satisfy 'rational' members of a committee tend to ascribe to a committee 'irrational' preferences between gambles, where rationality is defined in Chapter 3. Some of the reasons why this should be the case are given in the later sections of this chapter. The decision analyst should therefore, whenever possible, work for one client or be personally responsible to an organization, rather than take on the role of arbitor. Of course there are times when this is not possible. In these situations some help can still be given. A brief discussion of some of these techniques is given in Section 5.3.

## 5.2 INFLUENCE DIAGRAMS

### 5.2.1 Introducing influence diagrams

An influence diagram is a schematic representation of a decision problem. Unlike the decision tree, it represents the relationships between the component decision spaces and uncertain quantities or random variables rather than the relationships between each possible combination of decision and outcome that might occur. Therefore the influence diagram often provides a more compact representation of a problem than does a decision tree. This is especially true if the corresponding decision tree is symmetric in its branches. Because of the simplicity of an influence diagram, a practitioner will often prefer it to a decision tree. In addition it has the advantage over a decision tree of representing succinctly any conditional independence between variables. We shall show later that ideas of conditional independence are central to model building and the assessment of probabilities.

Unfortunately influence diagrams are no panacea. There is no unique influence diagram for any problem. Influence diagrams also tend to encourage confusion between causality and conditional independence. However, they are an important tool. Here is how they are constructed and used.

An influence diagram $I$ is a directed graph with nodes classified in one of three ways. The types of node are:

(a) a *chance node* $n(c)$ – represented by a circle;
(b) a *decision node* $n(d)$ – represented by a square box;
(c) a *value node* $n(v)$ – represented by a diamond shape.

A node $n_2$ is called a *direct successor* to a node $n_1$ if there is a directed arc from $n_1$ to $n_2$ in $I$. In this case $n_1$ is called an *direct predecessor* of $n_2$. Chance nodes label random variables/uncertain quantities. Decision nodes represent component decision spaces and value nodes represent the expected utility conditional on the values of its direct predecessing nodes. In our oil drilling example of Chapter 2, the chance nodes are:

$n(A)$ – labelling the presence or otherwise of oil in field $A$
$n(B)$ – labelling the presence or otherwise of oil in field $B$
$n(T)$ – labelling the results of any test we choose to do.

Chance node $n(c_1)$ must be a direct predecessor of chance node $n(c_2)$ if, and only if, the distribution of the random variable $X_2$ labelled by $n(c_2)$ is calculated conditional on the value of random variable $X_1$ labelled by $n(c_1)$ and $X_1$ and $X_2$ are not independent. Thus if in a decision problem you have chosen to express your client's joint mass function $p(x_1, x_2)$ of non-independent random variables $X_1$ and $X_2$ using the decomposition

$$p(x_1, x_2) = p(x_2 | x_1)p(x_1).$$

then this decomposition would be represented as in Fig. 5.1. If $X_1$ and $X_2$ were independent then no arc would connect $n(c_1)$ and $n(c_2)$.

More generally, let $n$ chance nodes $n(c_1), \ldots, n(c_n)$ represent random variables $X_1, \ldots, X_n$. The chance node $n(c_r)$ can be represented as having direct predecessors $n(c_{1_r}), \ldots, n(c_{k_r})$, $1 < 1_r, \ldots, k_r < r-1$ if and only if, conditional on the value of $X_{1_r}, \ldots, X_{k_r}$, $X_r$ is independent of $(X_1, \ldots, X_{r-1})$.

Provided $X_1, \ldots, X_n$ have a mass function which is non-zero everywhere in its argument, the definition can be expressed much more simply in terms of the form of the joint mass function. For example, the joint mass function

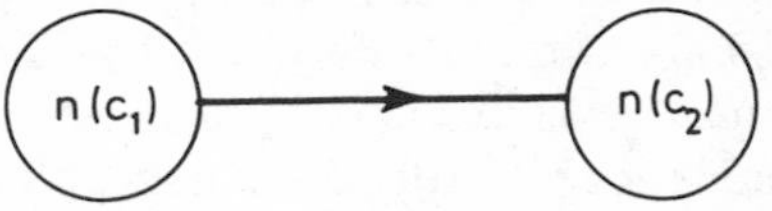


**Figure 5.1**

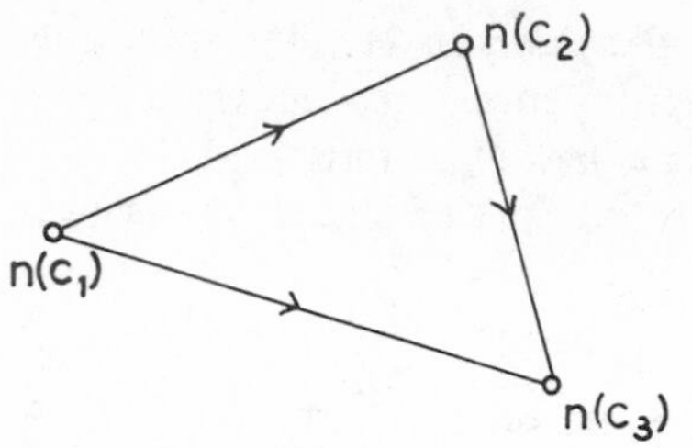


Figure 5.2

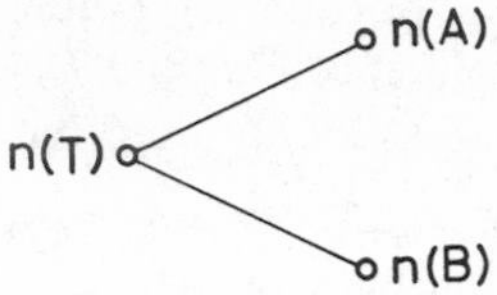


Figure 5.3

$p(x_1, x_2, x_3)$ of three random variables $X_1, X_2, X_3$ can be decomposed thus:

$$p(x_1, x_2, x_3) = p(x_3|x_2, x_1)p(x_2|x_1)p(x_1) \tag{5.1}$$

If node $n(c_i)$ represents random variable $X_i$, $1 \leqslant i \leqslant 3$, this relationship can be expressed as in Fig. 5.2, provided each of the conditional mass functions above are non-constant in all their arguments. Note that $n(c_3)$ has $n(c_1)$ and $n(c_2)$ as direct predecessors; $n(c_2)$ has $n(c_1)$ as a direct predecessor; and $n(c_1)$ has no direct predecessor. If, for example, $X_3$ and $X_2$ were independent conditional on $X_1$, $p(x_3|x_2,x_1)$ would be constant in $x_2$ and the arc in this diagram joining $n(c_2)$ to $n(c_3)$ need not be drawn.

The probabilistic relationship in our drilling example between the various elements of uncertainty can be represented by Fig. 5.3.

Here you are given the probabilities of the results of the test *conditional* on oil being present in field $A$ or $B$. On the other hand, the probabilities of oil in $A$ or oil in $B$ are given marginally and these events are assumed *a priori* independent. Hence no arc joins $n(A)$ and $n(B)$.

Note that the influence diagram representation of uncertain quantities is never unique. For example, you could always have decomposed the joint probability mass function given in (5.1) in a different way, e.g.

$$p(x_1, x_2, x_3) = p(x_1|x_2, x_3)p(x_2|x_3)p(x_3)$$

This would just have reversed the direction of our arcs in Fig. 5.2. However, there is usually a 'most useful' representation of uncertainty which exploits (conditional) independence within the system. Note here, that because of the usual rules of combining probabilities, no breakdown of the joint mass function can lead to a circuit in our chance nodes in $I$. So if a diagram admits a circuit it does not represent a decomposition of a joint mass function.

A decision node $n(d)$ is a direct predecessor of a chance node $n(c)$ if the distribution of the random variable labelled by $n(c)$ can depend on the choice of decision in the decision space labelled by $n(d)$. Thus, in our drilling example, the only uncertain quantity influenced by a decision your client takes is represented by the node $n(T)$. The particular decision $d_1$ which influences $n(T)$ is the initial one – whether to investigate one of the oilfields and if so

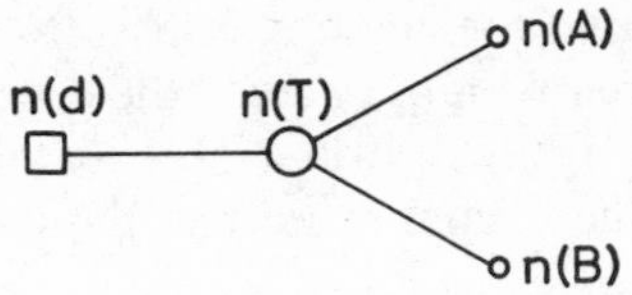


**Figure 5.4**

which one, field $A$ or field $B$. Clearly $d_1$ influences the results of your client's test so its associated node $n(d_1)$ should be connected by an arc to $n(T)$. On the other hand, $d_1$ has no effects on the marginal probabilities of oil in $A$ or oil in $B$ represented by nodes $n(A)$ and $n(B)$. So $n(d_1)$ is not connected to either of these nodes. Your influence diagram of this problem has now grown to look like Fig. 5.4.

A decision node $n(d_1)$ is a direct predecessor of another decision node $n(d_2)$ if and only if decision $d_2$ will be made *after* decision $d_1$ has been taken and the decision $d_1$ remembered when taking decision $d_2$. A chance node $n(c)$ is a direct predecessor of a decision node $n(d)$ if and only if the value of the uncertain quantity labelled by $n(c)$ will be known at the time the decision $d$, labelled by $n(d)$, is taken and might influence that decision.

A value node represents the client's expected utility given the values of its direct predecessors. It therefore represents a function of each of the decisions and possible outcomes of uncertain quantities labelled by its direct predecessor decision and chance nodes. So directed arcs enter a value node if the client's expected utility conditional on the problem as a whole, is functionally dependent on the values of the variables labelled by those directly predecessing nodes. We shall henceforth assume that all problems can be expressed by an influence diagram with *one* value node with no direct successors.

You can now draw the influence diagram of the whole drilling example of Chapter 2. Assuming the client wishes to maximize his expected pay-off, i.e. that his utility is linear in pay-off, then it can be drawn and represented by Fig. 5.5.

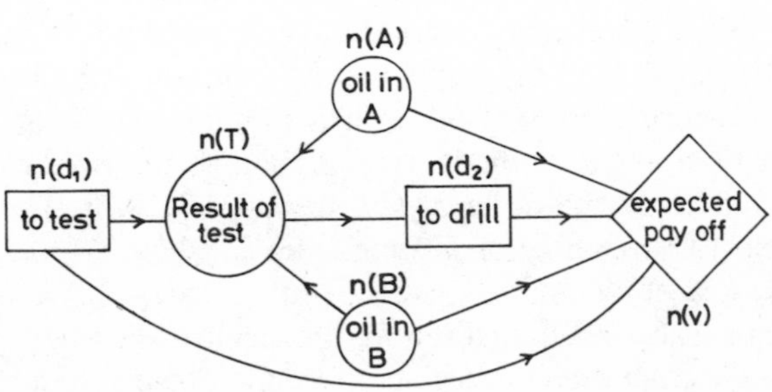


**Figure 5.5** Completed influence diagram

You have already drawn the relationship between $n(d_1)$, $n(A)$, $n(B)$ and $n(T)$ in Fig. 5.4. There are two more nodes: the decision node $n(d_2)$ labelling the decision $d_2$ whether or not to drill and if so, which field; and the value node $n(v)$ representing the client's expected pay-off. His final pay-off will depend on:

(a) $d_1$ (since if he chooses to test a site and then decides not to drill he will lose the cost of the investigation);
(c) events $A$ and $B$ of oil being found;
(c) $d_2$, the decision of which field, if either, to drill.

Hence $n(d_1)$, $n(d_2)$, $n(A)$ and $n(B)$ are all connected to $n(v)$. Decision $n(d_2)$ will be made solely in the light of any test he chooses to do.

So $d_2$ will depend only on the result, if any, of any test $T$ and $d_1$, the decision whether and what to test. Thus only two nodes, $n(T)$ and $n(d_1)$ are connected by an arc to $n(d_2)$. This completes our influence diagram. Note how much simpler this diagram is than the decision tree representation of this problem.

### 5.2.2* Evaluating an 'extensive form' influence diagram

We shall say that an influence diagram whose single value node has no direct successors is in *extensive form* if:

(a) it contains no (directed) cycles;
(b) the only node to have no direct successor is the value node;
(c) if any chance node $X$ is not a direct predecessor of any decision node $D$, then there is no directed path from $X$ to $D$;
(d) there is an ordering of the decision nodes $(D_1, \ldots, D_m)$ such that $D_r$ is a direct successor of $D_k$ where $k < r$ and $1 \leqslant r \leqslant m$.

If an influence diagram is in extensive form then its decision spaces and uncertain quantities are in an order required to evaluate the corresponding decision tree. Note that the drilling example is not yet in extensive form. In Appendix 2, Theorem A2.1, we show that if an influence diagram $I$ is in extensive form then there exists one node whose only direct successor is the value node. We shall use this result to construct an influence diagram $I_1$ equivalent to $I$ for the purposes of evaluating an optimal decision rule which is in extensive form and has one less node than the original diagram $I$.

Denote by $n(v)$ the value node of $I$. The node $n(v)$ will have an expected utility $\bar{G}$ which is a function of all variables labelled by nodes which are direct predecessors of $n(v)$.

If a chance node $n(1)$ labelling a random variable $X_1$ is a direct predecessor of $n(v)$ and has no other direct successors, you can absorb it into your client's influence diagram as follows. Since $n(v)$ is the only direct successor to $n(1)$

the value $x_1$ of $X_1$ influences nothing in the problem other than the final expected utility $\bar{G}$. So calculate $\bar{G}_1$, the expectation of $\bar{G}$ taken across $X_1$ for each of the possible values of the variables whose nodes directly precede $n(v)$. $\bar{G}_1$ is an explicit function of:

(i) variables labelled by all other direct predecessing nodes of $n(v)$ in $I$;
(ii) all variables labelled by direct predecessing nodes of $n(1)$.

To draw an influence diagram $I_1$ equivalent to $I$, delete node $n(1)$ from $I$ and replace $n(v)$ by $n_1(v)$ labelling the expected utility $\bar{G}_1$. By statements (i) and (ii) above, all arcs into $n(1)$ and $n(v)$ in $I$ need to be redirected into $n_1(v)$, other than the arc from $n(1)$ to $n(v)$ which is deleted. All other arcs which are in $I$ remain identical in $I_1$, no operation having been performed which affects these relationships.

On the other hand, if there is no chance node whose only successor is the value node, then by Theorem A2.1 this property must be held by a decision node. By condition (*d*) this decision node must be node $D_m$. Relabel this node $D_m$ by $n(1)$. By condition (d) all other decisions will be known when the decision from $D_m$ is taken. In Theorem A2.2, we show that if $I$ has no chance node whose only direct successor is $n(v)$ then every chance node in $I$ is a direct predecessor of $D_m$. So at the time a decision from $D_m$ is taken, all uncertainty labelled by the chance nodes in the diagram will be known. Therefore, given the direct predecessors of $n(1)$, $\bar{G}$ is a *deterministic function* of $d_m \in D_m$.

As before, you can therefore draw a new influence diagram $I_1$ representing the original problem which assumes the best decision in $D_m$ is inevitably taken. Given all the information available when a decision $d_m \in D_m$ is taken, calculate the decision $d_m^*$ which maximizes $\bar{G}$ as a function of this information. Let $\bar{G}_1$ denote $\bar{G}(d_m^*)$ and note that $\bar{G}_1$ is a function of (i) and (ii) mentioned above. Remove the node $n(1)$ from $I$ and replace $n(v)$ by $n_1(v)$ labelling $\bar{G}_1$. Since $\bar{G}_1$ is an explicit function of all variables mentioned in (i) and (ii), all arcs in $I$ into $n(v)$ and $n(1)$ are directed into $n_1(v)$ other than the arc from $n(1)$ to $n(v)$ which is deleted. As before, since this maximization is done conditional on other relevant variables, all arcs in $I$ not involving nodes $n(1)$ and $n(v)$ remain unchanged in $I_1$.

We have thus seen that any influence diagram $I$ with $n$ nodes in extensive form can be re-represented as an influence diagram $I_1$ defined above with one less node, i.e. $n-1$ nodes. It is proved in Theorem A2.3 that $I_1$ is itself in extensive form. It follows that you can use the same operations to find an influence diagram $I_2$ with $n-2$ nodes which is equivalent to $I_1$ and hence $I$ and which is also in extensive form. Proceeding inductively in this way you will eventually have constructed an influence diagram with only one node, a value node representing the expected pay-off of the problem given a sequence of optimal decisions is taken. Furthermore, the optimal decisions associated with node $n(r)$ in $I$ can be recorded when, under the operation described above, $n(r)$ is deleted to form the influence diagram $I_r$.

Thus, given that a diagram $I$ is in extensive form, you can use the algorithm above to evaluate your client's expected utility given that he acts optimally, and also use it to determine which actions attain this optimal expected utility.

One question that remains unanswered is 'Does such an algorithm always give the same result regardless of the order of selection in which you choose your direct successors in your reduction algorithm given above?' The answer to this question is 'Yes'. The reason is that different reductions of $I$ will give different but equivalent breakdowns into various conditional expectations of utility. The laws of probability ensure that all the breakdowns that use the algorithm above will give the same result.

### 5.2.3* Evaluating influence diagrams not in extensive form

So far we have only constructed an algorithm for finding a Bayes decision from an influence diagram in extensive form. However, usually an influence diagram $I$ can be rerepresented as an extensive form influence diagram if $I$ has one value node which has no direct successor.

If $I$ has a circuit, however, then it cannot be rerepresented by a simple form influence diagram. This is because either the dependences you have been given do not fully specify a breakdown of joint mass function in terms of conditional probabilities or you have an ambiguous representation of your problem. You will be asked to show this in Exercise 5.3 at the end of this chapter. If you find you have built such an influence diagram, either you need further probability information for complete specification of your problem or you have misrepresented it. In practice it is more usual for the latter case to pertain. The ambiguity in the diagram can often be resolved by either:

1. Expressing the problem in terms of a larger number of smaller decision spaces and uncertain quantities and breaking down uncertain quantities in a simpler way.
2. Introducing a temporal index on nodes in the diagram.

Unfortunately detailed discussion of such techniques is beyond the scope of this book (see Smith, 1987b).

On the other hand, it is easy to adjust an influence diagram $I$ so that it satisfies condition (b) of the extensive form definition. Check the decision and chance nodes of $I$, one by one, to see if they have a direct successor. If any decision or chance node $n$ does not have a direct successor delete it from your diagram together with all arcs into $n$. Keep repeating this procedure until the resulting influence diagram only contains decision and chance nodes with successors.

It is not difficult to see why this contracted graph is equivalent to $I$ for the purposes of identifying an optimal policy. Let $\bar{G}$ denote the conditional expected utility associated with the value node $n(v)$. First suppose that a decision space $D$ containing decision rules $d$ is represented by a decision node with no direct successor. Then, by the definition of an influence diagram this means:

(a) $\bar{G}$ is not an explicit function of $d \in D$ (since $n$ is not connected to $n(v)$);
(b) no other later decision is made in the light of $d \in D$ (since $n$ is not connected to another decision node);
(c) your client's choice of $d \in D$ does not influence the distribution of any uncertain quantity (since $n$ is not attached to a chance node).

Thus your client's expected utility across the whole problem cannot be affected by his choice of $d \in D$.

On the other hand, if $n$ is a chance node with no direct successor and $n$ labels a random variable $X$, you can conclude:

(a) $\bar{G}$ is not an explicit function of the value $x$ that $X$ takes (since $n$ is not connected to $n(v)$);
(b) no decision is made in the light of observing $X$ (since $n$ is not connected to any decision node);
(c) no probability distribution in the problem is given conditional on $X$ (since $n$ is not connected to any chance node).

It follows that the observed value $x$ of $X$ cannot influence the choice of optimal policy.

Condition (d) of the definition of extensive form will always hold for influence diagrams of problems when it is possible to make decisions in an order where your client does not forget what he has previously done. Most decision problems acted upon by a single decision-maker fall into this category.

Condition (c) is however commonly violated. Indeed our drilling example of Fig. 5.5 breaks this condition. However, this can be remedied by using an alternative but equivalent breakdown of the joint distribution of random variables in the problem into various conditional distributions which ensure that this condition is satisfied. For, by reversing conditioning, we reverse directions of arrows between nodes.

In our example, rather than using the original breakdown of the joint mass function of $(T, A, B)$

$$P(T, A, B) = P(T \mid A, B)P(A)P(B)$$

you can use the alternative breakdown

$$P(T, A, B) = P(A, B \mid T)P(T)$$

Notice that because your client tests either field $A$ or field $B$ but not both, and $A$ and $B$ are initially independent, either

$$P(A, B \mid T) = P(A \mid T)P(B)$$

or

$$P(A, B \mid T) = P(A)P(B \mid T)$$

In either case $A$ and $B$ will be independent conditional on $T$ so that

$$P(T, A, B) = P(A \mid T)P(B \mid T)P(T)$$

It follows that another influence diagram of his problem is one where the arcs between $A$ and $T$ and $B$ and $T$ are reversed. It is easy to check that this new influence diagram is in extensive form.

You should exercise great care when writing down alternative representations of a joint mass function of the random variables in your model. For example, in the above if $T$ were a test which depended on both $A$ and $B$ in some way then an extra arc would need to be drawn on to the diagram linking node $A$ with node $B$, because in general

$$P(A, B \mid T) \neq P(A \mid T)P(B \mid T)$$

even if $A$ and $B$ are marginally independent.

Algorithms for reversing arcs are given in Shachter (1984, 1986) and Smith (1986, 1987a)

### 5.2.4 Influence diagrams, causal diagrams and expert systems

We showed in Section 4.6 that it was often a good idea to construct probabilities about events by conditioning using causal connections. There is often an influence diagram which closely corresponds to a causal diagram of the problem. Causal diagrams can thus be used to help construct influence diagrams although the two types of diagram are not the same (see Exercise 5.4). Kim and Pearl (1983) discuss links between causality and influence diagrams.

An *expert system* is an algorithm which generates a 'best' diagnosis of the state of an individual on the basis of being given his symptoms. Typically these rules linking diagnoses and symptoms are built on an extremely large bank of observations of both symptoms and states of 'similar' individuals. Both Spiegelhalter (1986a) and Pearl (1982) suggest using influence diagrams to help in the construction of expert systems.

A good introduction to the properties and applications of influence diagrams is given in Howard and Matheson (1981) and Shachter (1984, 1986). An alternative diagram to represent relationships between uncertain events is given in Darroch *et al.* (1980). Comparisons using an example are given in Spiegelhalter (1986a, b).

## 5.3 THE COMMITTEE AS A DECISION-MAKER

### 5.3.1 Introduction

Suppose that a committee, rather than a single individual, needs to make decisions about a set of problems. Assume that each member of that committee has agreed to a set of his preferences which are rational in the sense given in Chapters 3 and 4. In particular, each member has his own expected utility on the various group decisions open to the committee. Can we ensure that this committee or *decision centre* has preferences over its options, which are derived from the stated preferences of its members, and which also satisfy the utility rules of Chapter 3? And, if this is possible, how should we derive a utility for the decision centre as a function of the utilities in the committee? These are the questions we shall address in the following sections.

### 5.3.2 Democracy, consensus and irrationality

Two commonly used ways of combining preferences are the democratic vote and consensus policy. In a *democratic vote* a distribution of rewards $P_1$ is preferred to $P_2$ if more individuals on the committee prefer $P_1$ to $P_2$. You might be surprised to learn that the democratic voting system combines the preferences of the members in the decision centre in such a way that the preferences of that decision centre will in general violate the utility rules of Chapter 3.

To illustrate this, consider a committee of three members $M_1$, $M_2$ and $M_3$ with respective expected utilities $\bar{u}_1, \bar{u}_2, \bar{u}_3$ on the following distributions of reward $P_1, P_2, P_3$. Suppose

$$\bar{u}_1(P_1) < \bar{u}_1(P_2) < \bar{u}_1(P_3)$$
$$\bar{u}_2(P_2) < \bar{u}_2(P_3) < \bar{u}_2(P_1)$$
$$\bar{u}_3(P_3) < \bar{u}_3(P_1) < \bar{u}_3(P_2)$$

If the committee decides on its preferences by a democratic vote across pairs of options, then it should prefer $P_2$ to $P_1$, because both $M_1$ and $M_3$ prefer $P_2$ to $P_1$; it should prefer $P_3$ to $P_2$, because $M_1$ and $M_2$ have this preference; and it should prefer $P_1$ to $P_3$, because this agrees with both $M_2$'s and $M_3$'s preferences. Suppose there existed a group utility function $u$. Then these preferences would imply that

$$\bar{u}(P_2) > \bar{u}(P_1) \tag{5.2}$$

$$\bar{u}(P_3) > \bar{u}(P_2) \tag{5.3}$$

$$\bar{u}(P_1) > \bar{u}(P_3) \tag{5.4}$$

Clearly these inequalities cannot hold simultaneously since (5.2) and (5.3) imply that

$$\bar{u}(P_3) > \bar{u}(P_1)$$

which contradicts (5.4). So no group utility $u$ can exist. In fact the committee appears from the outside quite illogical, preferring $P_2$ to $P_1$, $P_3$ to $P_2$, but $P_1$ to $P_3$.

A decision centre run on the idea of *consensus* can also exhibit preferences which from the outside seem illogical. Suppose a decision centre will only accept a gamble if *all* members on the committee consider the gamble preferable to not taking the gamble. Otherwise the committee rejects the gamble in favour of doing nothing. Again assume that each member $M_i$ acts to maximize his expected utility $\bar{u}_i$ and without loss of generality assume that $\bar{u}_i(P_0) = 0$, $1 \leqslant i \leqslant m$, where $P_0$ is the reward distribution of obtaining nothing with certainty, the consequence of not accepting a gamble.

Decision centres run on the idea of consensus are quite common. Courts of law will often only convict if *all* members of the jury believe that in 'all probability' the defendant perpetrated the offence. We shall explore the next example in more detail in later sections.

A decision centre is comprised of representatives of several insurance companies. The centre may be offered the option of insuring an oil tanker for a certain yearly permium. Because claims will usually be very large if a tanker is damaged, representatives will agree beforehand that, in the event of an accident, each will make prearranged contributions to the claim settlement. In return each company obtains a prearranged portion of the yearly premium paid by the shipping company. Clearly here the decision centre will only be able to accept the gamble of insuring the tanker if they can all agree a mutually acceptable share-out of premiums and risk. Hence consensus of companies is necessary before a gamble can be accepted.

To illustrate how the consensus policy of combining preferences need not give rise to a group utility $u$, consider the simplest case when our committee consists of just two members $M_1$ and $M_2$, with respective utility functions $u_1$ and $u_2$. Assume, without loss of generality, that $u_1$ and $u_2$ are scaled so that $\bar{u}_1(P_0) = \bar{u}_2(P_0) = \bar{u}(P_0) = 0$, where $P_0$ represents the reward distribution that assigns each member (and hence the group as a whole) nothing with certainty and where $u$ is a candidate for the utility function of the group. Suppose there exist two reward distributions $P_1$ and $P_2$ satisfying

$$u_1(P_1) > 0, \quad \bar{u}_2(P_1) < 0 \tag{5.5}$$

$$\bar{u}_1(P_2) < 0, \quad \bar{u}_2(P_2) > 0 \tag{5.6}$$

$$\bar{u}_1(P_1) + \bar{u}_1(P_2) > 0 \tag{5.7}$$

$$\bar{u}_2(P_1) + \bar{u}_2(P_2) > 0 \tag{5.8}$$

Using the consensus policy, by inequalities (5.5) and (5.6) neither gamble $P_1$

nor $P_2$ should be accepted. So our candidate group utility $u$ must satisfy

$$\bar{u}(P_1) < 0, \quad \bar{u}(P_2) < 0 \tag{5.9}$$

Now consider the gamble $P$ for which $P_1$ will be offered with a probability $\frac{1}{2}$ or $P_2$ will be offered with probability $\frac{1}{2}$, i.e. $P = \frac{1}{2}P_1 + \frac{1}{2}P_2$. Then, using the linearity of expectation on $M_1$ and $M_2$'s utility functions,

$$\bar{u}_i(P) = \tfrac{1}{2}\bar{u}_i(P_1) + \tfrac{1}{2}\bar{u}_i(P_2) > 0, \qquad i = 1, 2$$

by inequalities (5.7) and (5.8). Thus both $M_1$ and $M_2$ should prefer the gamble $P$ to status quo. Hence, by the consensus rule we must have that

$$\bar{u}(P) > 0$$

But, by the properties of expectation used on $u$,

$$\bar{u}(P) = \tfrac{1}{2}\bar{u}(P_1) + \tfrac{1}{2}\bar{u}(P_2) < 0$$

by inequalities (5.9). Thus we have obtained a contradiction. No utility function $u$ can express the group's preferences over the reward distributions given above.

We can thus conclude that if you require your committee to exhibit rational preferences then you need to combine the beliefs and utilities of its members in a rather less straightforward way than by using consensus between members or by holding a democratic vote.

### 5.3.3 Pareto optimality and group utilities based on preferences of the members of a group

I have shown that two commonly used ways of combining beliefs of members of a group lead to group preferences which do not obey the utility assumptions of Chapter 3. In this section I shall try to derive an expected utility for the group as a function of the preferences and beliefs of its members.

Suppose the group has to make a decision $d$ in a set of alternatives $D$. Suppose that each of the $m$ members $M_i$ of the group has beliefs about the outcomes influencing the rewards for the group given the various decisions $d$ and a utility on those rewards respectively given by $P_i(d)$ and $u_i(d)$. Assuming that $M_i$ is rational, he will want to maximize his expected utility $\bar{u}_i(d)$, $d \in D$. Furthermore, he will prefer $d_1$ to $d_2$ if and only if

$$\bar{u}_i(d_1) > \bar{u}_i(d_2)$$

It is at first sight reasonable to assume that our group expected utility $\bar{u}$, which determines the preferences of the group, depends only on $\bar{u}_i$, $1 \leqslant i \leqslant m$, the functions determining the preferences of the individual members of the group. Technically this condition can be written

$$\bar{u} = f(\bar{u}_1, \bar{u}_2, \ldots, \bar{u}_m) \tag{5.10}$$

for some fixed function $f$ not depending on $\{u_i(d), P_i(d), 1 \leqslant i \leqslant m, d \in D\}$.

Our problem is now to determine the form the function $f$ must take. Now clearly if all members agree that $d_1$ is at least as good as $d_2$ and at least one member prefers $d_1$ to $d_2$, it would be unnatural not to assume that the group would prefer $d_1$ to $d_2$. Technically this can be stated as follows. For all $d_1, d_2 \in D$, such that $\bar{u}_i(d_1) \geqslant \bar{u}_i(d_2)$, $1 \leqslant i \leqslant m$, with strict inequality for some $i$,

$$\bar{u}(d_1) > \bar{u}(d_2) \tag{5.11}$$

This condition is called the *Pareto optimality condition.* A decision $d_2 \in D$ is called *Pareto optimal* if there exists no other decision $d_1 \in D$ satisfying inequality (5.11). In general there are many Pareto optimal decisions for a single problem.

We next assume that $D$ is *convex*, i.e. that if $d_1$, $d_2 \in D$ then

$$d_3(\alpha) = \alpha d_1 + (1 - \alpha) d_2 \tag{5.12}$$

is also in $D$ for all values $0 \leqslant \alpha \leqslant 1$. The decision $d_3(\alpha)$ denotes the decision rule which applies decision $d_1$ with probability $\alpha$ and $d_2$ with probability $1 - \alpha$. A decision space $D$ can always be embedded in a convex space $D^*$. For example, we can assume that for all pairs $d_1$, $d_2 \in D$ the committee can be offered the further option $d_3(\alpha)$ of agreeing to abide by the toss of an unfair coin having probability $\alpha$ of heads occurring. The committee will pursue option $d_1$ if a head results from the toss and otherwise pursue option $d_2$.

We finally assume that no single member or group of members dictates the preferences of the group. Explicitly we say that a proper subset $I_1$ of members in the group $I$ *dictates* the preferences of $I$ if for all decisions $d_i$, $d_j \in D$,

$$\bar{u}(d_i) > \bar{u}(d_j)$$

wherever

$$\bar{u}_k(d_i) \geqslant \bar{u}_k(d_j) \text{ for all } M_k \in I_1 \text{ with strict inequality for some } M_k.$$

If there did exist a group of members $I_1$ dictating the preferences of the group, it can be shown that the preferences of members in $I$ but outside $I_1$ would only be taken into account if the members in $I_1$ found two decisions equally preferable. In this sense the preferences of members in $I_1$ is considered infinitely more important than those of members outside $I_1$. To keep members' preferences on a comparable scale (see analogous Rule 4 of Chapter 3) we exclude functions $f$ which cause any group $I_1$ to dictate preferences.

It can be shown that a decision $d$ is Pareto optimal, when $D$ is convex and $u$ does not admit a dictating subset of members, if and only if

$$\bar{u}(d) = \sum_{i=1}^{m} \lambda_i \bar{u}_i(d) \tag{5.13}$$

for some values $\lambda_i > 0$, $1 \leqslant i \leqslant m$, where $\sum_{i=1}^{m} \lambda_i = 1$. We shall call any such decision $d$ *linear Pareto optimal.* You are lead through a proof of this result when $m = 2$ in Exercise 5.5. In addition, if $u_1, \ldots, u_m$ have been linearly translated so that $\bar{u}_i(d) \geqslant 0$ holds for all $d \in D$, $1 \leqslant i \leqslant m$, $d$ is linear Pareto optimal if and only if

$$\bar{u}^*(d) = \sum_{i=1}^{m} \bar{u}_i^{a_i} \tag{5.14}$$

for some values $a_i > 0$, where $\sum_{i=1}^{m} a_i = 1$.

### 5.3.4 When a committee can agree the distribution of rewards for all decisions

Let members $M_1, \ldots, M_m$ have respective utility functions $u_1, \ldots, u_m$ and let each agree about the appropriate distribution $P(d)$ of rewards $r$ for any decision $d \in D$ open to the group, where $D$ is convex. The distributions $P(d)$ can then be thought of as the group's probability assessments of $d \in D$. Furthermore, the function $\bar{u}(d)$ defined in equation (5.13) now satisfies

$$\bar{u}(d) = E\{u(d, r)\} \tag{5.15}$$

where this expectation is taken across $P(d)$ and

$$u(d) = \sum_{i=1}^{m} \lambda_i u_i(d, r), \qquad \sum_{i=1}^{m} \lambda_i = 1, \qquad \lambda_i \geqslant 0, 1 \leqslant i \leqslant m \tag{5.16}$$

Clearly now, the committee acts rationally, in the sense of Chapter 3, if we let the function $u$, defined in equation (5.16), be a *group utility function* for some fixed value of $\boldsymbol{\lambda} = (\lambda_1, \ldots, \lambda_n)$ not depending on $P(d)$. For then the group chooses the decision $d$ which maximizes its expected utility with respect to the distribution $P(d)$ and the utility function $u$. And by equation (5.13) such an optimal decision will also be linear Pareto optimal. So it will not be possible for the committee to find another decision which is more preferable to *all* members of the committee than the one chosen in the way described above.

Of course there are many functions $u$ satisfying equation (5.16) so we still need to resolve the contentious issue of picking appropriate values of $\boldsymbol{\lambda}$. Typically, increasing $\lambda_i$ relative to $\lambda_j$ increases the influence of the preferences of member $M_i$ on the group relative to the preferences of $M_j$. One way of trying to numerically fix $\boldsymbol{\lambda}$ using ideas from Section 3.6 is given in Exercise 5.8.

### 5.3.5 The problem of how to divide pay-offs among members of a syndicate

We consider here a simple but interesting example of a group decision problem illustrated by the influence diagram in Fig. 5.6. Each member $M_i$ has a utility function $u_i$ which is a function only of $r_i$, his monetary pay-off

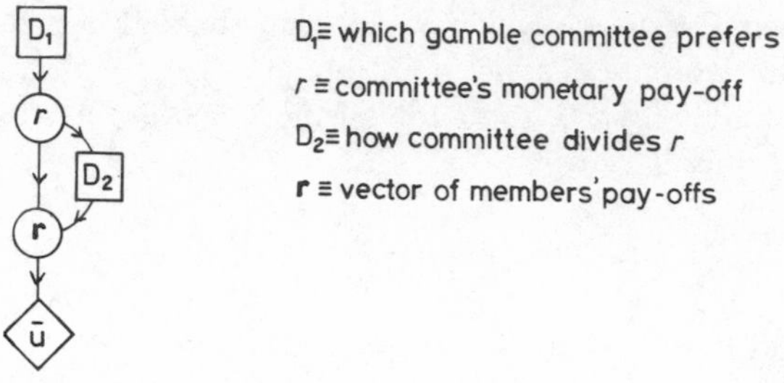


**Figure 5.6** An influence diagram of a simple group decision problem

from taking a particular gamble. Let $\mathbf{r} = (r_1, r_2, \ldots, r_m)$. On accepting a bet $d_1 \in D_1$ all the members agree that the group's reward $r$ will have distribution $P(d_1)$. The committee needs to agree a contingency decision $d_2 \in D_2$ which divides a pay-off $r$ in such a way that $M_1$ receives pay-off $r_i$, $1 \leqslant i \leqslant m$, $\sum_{i=1}^{m} r_i = r$.

In this section, I will illustrate how a committee can find its Pareto optimal partitions of rewards explicitly given $u_1, \ldots, u_m$. We shall also see whether the behaviour of the committee appears from the outside to be acting rationally.

I showed in Section 5.3.4 that any linear Pareto optimal partitions or reward will maximize the expectation of $u = \sum_{i=1}^{m} \lambda_i u_i(d, \mathbf{r})$ for some value of $\boldsymbol{\lambda}$ such that $\lambda_i > 0$, $1 \leqslant i \leqslant m$ and $\sum_{i=1}^{m} \lambda_i = 1$. Now the decision $d_2 \in D_2$ of how to partition the committee's pay-off among its members is made in the light of observing that group's pay-off $r$. Hence, given a betting scheme $d_1 \in D_1$, a partition $d_2 \in D_2$ will maximize the expectation of $u$, if and only if it is a function of $r$ which maximizes $u(d_2, r)$ for a set of rewards $r$ which will occur with probability one after decision $d_1$ is taken.

Thus, in this problem, the Pareto optimal decisions can be identified without reference to the distribution $P(d_1)$. This eases identification considerably. Here is an example of how this is achieved when $M_1, \ldots, M_m$ have constant risk-averse utilities (introduced in Exercise 3.7).

*Example 5.1*

The group chooses a decision $d_1 \in D_1$ and receives a reward $r$. Suppose that the group pay-off $r$ is divided so that member $M_i$ receives an amount $r_i$, $1 \leqslant i \leqslant m$. Then $M_i$'s utility $u_i$ is assumed to be of the form

$$u_i(\mathbf{r}) = 1 - \exp\{-\theta_i^{-1} r_i\} \qquad \theta_i > 0, 1 \leqslant i \leqslant m \tag{5.17}$$

where $\mathbf{r} = (r_1, \ldots, r_m)$ and $\sum_{i=1}^{m} r_i = r$.

By the arguments given above, any linear Pareto optimal partition $d_2$ must maximise $\sum_{i=1}^{m} \lambda_i u_i(\mathbf{r})$ under the constraint $\sum_{i=1}^{m} r_i = r$, for some fixed value of $\boldsymbol{\lambda}$. By the method of Lagrange multipliers (e.g. see Craven, 1981) this is equivalent to maximizing

$$u^*(\mathbf{r}) = \sum_{i=1}^{m} \lambda_i u_i(\mathbf{r}) + A\left(\sum_{i=1}^{m} r_i - r\right)$$

where $A$ is chosen so that the constraint $\sum_{i=1}^{m} r_i = r$ is satisfied.
Now,

$$\frac{\partial u^*(\mathbf{r})}{\partial r_i} = 0 \Leftrightarrow -\lambda_i \theta_i^{-1} \exp\{-\theta_i^{-1} r_i\} + A = 0$$

When $\lambda_i \neq 0$, $1 \leqslant i \leqslant m$, this implies that

$$r_i = (\log A)\theta_i + \rho_i \qquad \text{where } \rho_i = (\log \lambda_i - \log \theta_i)\theta_i \tag{5.18}$$

Now $A$ must be chosen so that $\sum_{i=1}^{m} r_i = r$, which by summing the $m$ equations above implies

$$r = (\log A) \sum_{i=1}^{m} \theta_i + \sum_{i=1}^{m} \rho_i$$

i.e. $\log A = \theta^{-1}(r - \rho)$ where we define $\rho = \sum_{i=1}^{m} \rho_i$ and $\theta = \sum_{i=1}^{m} \theta_i$. Substituting for $\log A$ into equation (5.18) gives us that all linear Pareto optimal partitions with $\lambda_i > 0$, $1 \leqslant i \leqslant m$, are of the form which gives a reward $r_i$ to $M_i$ which satisfies

$$r_i = \psi_i r - \tau_i \tag{5.19}$$

where $\quad \psi_i = \theta_i/\theta \quad \theta = \sum_{i=1}^{m} \theta_i \quad$ and $\quad \tau_i = \rho\psi_i - \rho_i, 1 \leqslant i \leqslant m.$

The reader might like to check that the Pareto optimal partition corresponding to vector $\boldsymbol{\lambda} = (\lambda_1, \ldots, \lambda_n)$ with one or more zero elements, takes on the form given in equation (5.19) except that those members $M_i$ for which $\lambda_i = 0$ are excluded from the group and given a zero pay-off with probability 1. Clearly, randomizing between two such partitions gives a partition of the same form but with $\tau_i$ randomized.

Thus if $M_1, \ldots, M_m$ all have constant risk-averse utilities we have shown that the only partitions which are candidates for 'optimal' ones demand that member $M_i$ pays an amount $\tau_i$ (determined by the committee's agreed choice of $\boldsymbol{\lambda}$) and receives a proportion $\psi_i$ of the group's winnings $r$, $1 \leqslant i \leqslant m$. Note that this proportion $\psi_i$ depends only on $M_i$'s utility function and not on $\boldsymbol{\lambda}$. Thus once it is agreed that only linear Pareto optimal decisions should be taken, the only indeterminacy in the group's choice of decision concerns the amount $\tau_i$ that member $M_i$ should pay (or receive) to participate in the bet.

Now let us turn to the problem of how such a committee appears to behave from an outsider's point of view. Typically the outsider may well know the distribution $P(d_1)$ of the rewards given a group decision $d_1$ but will rarely know the mechanism by which that reward $r$ will be divided among its members. Consequently, the outsider has only partial information ($r$ rather than $\mathbf{r}$) about the reward space of the group. We can now ask the question: Does the committee act as if it were a single decision-maker whose pay-off was $r$?

It is not surprising that the answer to this question for general forms of utility function is 'No'. An example illustrating this is given in Exercise 5.9. However, in the example above, when all members are constant risk-averse we get the somewhat more surprising answer of 'Yes'. For when $\lambda_i > 0$, $1 \leqslant i \leqslant m$, for *any* possible values of $\tau_1, \ldots, \tau_m$ defined in equation (5.19), $M_i$'s utility for the *group* to receive $r$ is

$$\begin{aligned} u_i(\mathbf{r}) &= 1 - \exp\{-\theta_1^{-1}(\psi_i r - \tau_i)\} \\ &= 1 - \exp\{-\theta^{-1} r\} \exp\{\theta_i^{-1} \tau_i\} \end{aligned}$$

It follows that all members' utilities on what the group receives are increasing linear functions of each other and so are equivalent (see Chapter 3). Whatever the final agreement on the side payments $(\tau_1, \ldots, \tau_m)$ is, under the agreement that the committee should only choose linear Pareto optimal decisions, every member will have a utility on group pay-off $r$ which is equivalent to

$$u^*(r) = 1 - \exp\{\theta^{-1} r\} \qquad \text{where } \theta = \sum_{i=1}^{m} \theta_i \tag{5.20}$$

In particular, for any distribution $P(d_1)$ of rewards $r$, each member's expected utility will be proportional to $u^*(r)$. So the committee acts as if it is maximizing the expectation of $u^*(r)$ and appears to satisfy, from the outside, the assumptions governing rational choice.

Several practical group decision problems fall into the category of problems discussed in this section. For example, the committee of representatives of insurance companies given in an example in Section 5.3.2 divide their rewards after receiving (or not receiving) an accident claim on an insured item. Another example of risk sharing for oil wildcatters is given in Raiffa (1968).

### 5.3.6 The general problem of finding a group utility function

There is now considerable evidence that it is not possible to 'sensibly' assign a utility $u$ and a distribution $P$ to a group so that all Bayes decisions are Pareto optimal unless:

(a) one member dictates the preferences of the group; or
(b) all members agree on their probability assessments (see Section 5.3.4); or
(c) all members have equivalent utility functions; or
(d) the class of problems considered is severely restricted.

The term 'sensible' above is variously defined by authors (Bacarach, 1975; Arrow, 1951; West, 1984; Zidek, 1983). For a discussion of the problems that arise see French (1985). Two popular ways of combining utilities is to use equation (5.13) with constant $\lambda$'s or equation (5.14) with constant $a_i$. These are called respectively the *linear pool u* and *logarithmic pool u**. Each pool defines a preference ordering which satisfies some, but not all, of the utility axioms. We showed in Section 5.3.4 that in the case of all members agreeing on the distribution $P$ of uncertain quantities, the linear pool *does* act rationally. The group acts then to maximize the expected utility of $(u_G, P_G)$ where $P_G$ is the agreed distribution and $u_G = \sum_{i=1}^{m} \lambda_i u_i$. Similarly, if the utilities $u_1, \ldots, u_m$ are identical and equal to $u$, say, then $(u_G, P) = (u, P_G)$, where $P_G = \sum_{i=1}^{m} \lambda_i P_i$ where $P_i$ is the distribution function of the $i$th member. When $u_i$ and $P_i$ are allowed to be different, however, then sometimes we cannot find an appropriate pair $(u_G, P_G)$ such that $E(u_G(d)) = u^*(d)$ for all $d \in D$.

Some authors have suggested that it is more important that a group chooses decisions which are *fair* to its members rather than choose decisions which appear logical to an outsider. Nash (1950) argues that choosing a decision which maximizes the value of the logarithmic pool given by equation (5.14) with $a_i = m^{-1}, 1 \leqslant i \leqslant m$, gives a decision which is 'fair' to all its members.

### 5.3.7 Some concluding remarks about group decisions

We have shown that, whereas committees which follow commonly used rules to combine preferences can appear to act illogically, it is possible to combine preferences to be rational provided that members of a group all agree on their probability assessments. On the other hand, we have also referenced other results which suggest that when members can agree on neither their utility nor their probability assessments then the group cannot guarantee that it will satisfy the utility assumptions and at the same time satisfy the Pareto optimality condition (that no group-optimal decision is bettered by another in the view of all members of the committee).

In fact this last result should not be too surprising. Condition (5.10) implies that a combination rule should take no account of why members have their idiosyncratic beliefs and preferences. Together with Pareto optimality condition (5.10) also forces us to assume that each member's utility is only a function of his own gain. It disregards, in quite an unrealistic way, the group dynamics, ignoring, for example, the competition and altruism between subgroups of the constituent members. For example, Gibbard (1973) shows that, given any combination rule, it is always in the interest of at least one member to misrepresent his utilities and probabilities. In fact these criticisms also apply when members agree on their probability statements as discussed in Section 5.3.4.

In my opinion, the only way to resolve this problem is for the decision analyst to model the individual statements made by the members of the group. This involves the decision analyst treating each member's utility and probability statements as data which she uses to come to an optimal decision on behalf of the group. This process involves considerable expertise and is beyond the scope of this small text (see, for example, Lindley, 1985a).

For good reviews of all these problems see French (1985, 86) and Genest and Zidek (1986).

## EXERCISES

5.1 Draw influence diagrams to represent the decision problems of Examples 1.1 and 1.2.

5.2 Evaluate the influence diagram which represents the drilling problem of Chapter 2.

5.3 Suppose an influence diagram exhibits a directed cycle. We have shown that if that cycle involves only chance nodes then a full probability distribution breakdown has not been given.

(i) Show that if two decision nodes $n_1$ and $n_2$, representing respective decision spaces $D_1$ and $D_2$, lie on this cycle then $d_1 \in D_1$ is taken in the light of knowing $d_2 \in D_2$, and conversely $d_2 \in D_2$ is taken in the light of $d_1 \in D_1$.

(ii) Show that if this cycle contains exactly one decision node $n$, labelling decisions $d \in D$, then $d$ must be taken in the light of the observed value of some uncertain quantity whose distribution is influenced by $d$.

5.4 Causal diagrams were introduced in Section 4.6. Although it can be argued that the chosen representation of a problem by an influence diagram should coincide with a causal diagram as far as possible, the two types of representation are quite distinct.

(i) One of two computers $C_1$ and $C_2$ is chosen at random (an event represented by node $n_1$) to produce a random number $N$ in $[0, 1]$ (an event represented by node $n_2$). Show than $n_1$ connects to $n_2$ in the causal diagram but not in any influence diagram.

(ii) It has been noted that the sales of washing machines (an uncertain quantity represented by node $n_1$) is highly correlated with crime figures for that year (an uncertain quantity represented by node $n_2$). Show that $n_1$ and $n_2$ should be connected in the influence diagram but not in a causal diagram.

5.5* A group has two members $M_1$ and $M_2$ with respective expected utilities $\bar{u}_1(d)$ and $\bar{u}_2(d)$ over the space of possible decisions $d \in D$. Say $d^* \in V(\lambda)$ if $\lambda\bar{u}_1(d) + (1-\lambda)\bar{u}_2(d)$ attains its maximum value at $d^*$.

(a) Show that if $d^* \in V(\lambda)$ for some value $0 < \lambda < 1$, then it is a Pareto optimal decision.

(b) Now suppose $d'$ is some Pareto optimal decision in $D$ which is not in $V(\lambda)$ for any value of $0 < \lambda < 1$. Show that:

(i) $I \cup J = (0, 1)$, where

$I = \{\lambda \in (0, 1)$: there exists a $d^* \in V(\lambda)$ with $\bar{u}_1(d^*) > \bar{u}_1(d')$ and $\bar{u}_2(d^*) < \bar{u}_2(d')\}$

$J = \{\lambda \in (0, 1)$: there exists a $d^* \in V(\lambda)$ with $\bar{u}_1(d^*) < \bar{u}_1(d')$ and $\bar{u}_2(d^*) > \bar{u}_2(d')\}$

(ii) both $I$ and $J$ are open subsets of $(0, 1)$

(iii) if $D$ is convex (see Section 5.3.3) then $I$ and $J$ are disjoint. Hence, or otherwise, show that if $D$ is convex, any Pareto decision which lies in no $V(\lambda)$, $0 < \lambda < 1$, must be a Bayes decision for either $M_1$ or $M_2$ (i.e. that either $M_1$ or $M_2$ dictates the preferences of the group).

5.6 Two members $M_1$ and $M_2$ of a group have utility functions given by

$$u_1(x) = \log(b + x)$$
$$u_2(x) = \log(c + x) \qquad b, c > 0.$$

respectively. They always agree on probabilities of uncertain events and agree not to use decisions which are Bayes for one member but not the other. Use the result of Exercise 5.5 above to find the set of such Pareto optimal partitions of rewards between $M_1$ and $M_2$. Show that if $M_1$ and $M_2$ act in this way, then the group appears to act as if using a single utility function, and find this utility function explicitly.

5.7* Two sub-managers $P_1$ and $P_2$ have an option of jointly publishing their respective forecasts $d_1, d_2$ of the profits $\theta_1, \theta_2$ (respectively) they will make in the next year. However, their firm has already forecast that their joint profit will be $C$ and they are constrained to make $d_1 + d_2 = C$. $P_i$ believes his own profits $\theta_i$ have distribution with mean $\mu_i$ and variance $\sigma^2$, $i = 1, 2$. $P_i$ has utility function

$$u_i(d_i, \theta_i) = R - (d_i - \theta_i)^2, \quad R > 0, \quad i = 1, 2$$

and each manager has the option not to publish with utility value 0. Find the forecast decision pair $(d_1, d_2)$ corresponding to Nash's solution and interpret this result. Now suppose $\theta_1$ and $\theta_2$ have normal distribution with means and variance above and that $P_i$ uses instead the utility function

$$v_i(d_i, \theta_i) = \exp - \{\tfrac{1}{2}Q^{-1}(d_i - \theta_i)^2\}, \quad i = 1, 2$$

Show that the decision pair corresponding to Nash's solution is the same as for the first pair of utility functions provided that randomized decisions are not permitted. Show however, that the decision rule which lets $(d_1, d_2) = (\mu_1, C - \mu_1)$ with probability $\frac{1}{2}$ and $(d_1, d_2) = (C - \mu_2, \mu_2)$ with probability $\frac{1}{2}$ dominates Nash's pair if $|C - (\mu_1 + \mu_2)|$ is large enough. Why is the original Nash's solution dominated in this way?

5.8 A committee of $m$ members $M_1, \ldots, M_m$ have respective utility functions $u_1, \ldots, u_m$ and each share the same probability assessment $P(d)$ of rewards $r$ for all options $d \in D$. They need to fix $\lambda_1, \ldots, \lambda_m$ in their joint utility function

$$u = \sum_{i=1}^{m} \lambda_i u_i(d, r) \qquad \sum_{i=1}^{m} \lambda_i = 1$$

Let $M_1$ be a most influential member in the committee. Without loss of generality, scale $u_i(d, r)$ so that $\inf u_i(d, r) = 0$ and $\sup u_i(d, r) = 1$, $1 \leqslant i \leqslant m$.

Let $\delta_j$ be a (possibly hypothetical) option in which $M_j$ obtains utility 1 and all other members obtain utility $\frac{1}{2}$, $1 \leqslant j \leqslant m$. Let $\delta_0$ be a decision in which each member attains a utility $\frac{1}{2}$ and define $\alpha_j$ to be that probability

$$d(\alpha_j) = \alpha_j \delta_1 + (1 - \alpha_j)\delta_0 = \delta_j \qquad 2 < j < m$$

Show that if the group is coherent,

$$\lambda_1 = \left(1 + \sum_{j=2}^{m} \alpha_j\right)^{-1}$$

$$\lambda_j = \alpha_j \lambda_1 \qquad 2 \leqslant j \leqslant m.$$

5.9 (*After Raiffa*, 1968) A group has two members $M_1$ and $M_2$ having respective utility functions $u_1$ and $u_2$ given by

$$u_1(x) = \begin{cases} (3/2)x + 50 & -300 < x < -100 \\ x & -100 < x < 200 \\ (1/5)x + 160 & 200 < x < 300 \end{cases}$$

$$u_2(x) = \begin{cases} 2x + 200 & -300 < x < -200 \\ x & -200 < x < 100 \\ (3/5)x + 40 & 100 < x < 300 \end{cases}$$

The group is offered three betting schemes $d_1, d_2$ and $d_3$. Under $d_1$ two independent fair coins $c_1$ and $c_2$ will be tossed. If two heads occur the group will be paid \$300; if $c_1$ is a head and $c_2$ a tail the group will earn \$100; if $c_1$ is a tail and $c_2$ a head the group will lose \$100; and if two tails are tossed the group will lose \$300. Under $d_2$ a fair coin will be tossed and if the result is a head the group will gain \$100, if it is tails they will lose \$100. Under $d_3$ a fair coin will be tossed and the group receives \$300 if a head results and loses \$300 if a tail results. Show:

(i) that there is no partition of rewards which makes it preferable for both players to accept the bet rather than not, for either lottery $d_1$ or $d_2$.

(ii) that for both $d_1$ and $d_2$ there exist partitions of rewards such that both players are indifferent between accepting or not accepting the lottery.

(iii) that at least one player will want to veto any partition of rewards for $d_3$, so that $d_3$ cannot be accepted by the group.

If a group utility $u$ exists show that $\bar{u}(d_1) = \bar{u}(d_2) = 0$ and $\bar{u}(d_3) < 0$. By noting that the distribution of group reward from accepting the lottery $d_1$ is the same as the distribution of the lottery $d_4$ which chooses lottery $d_2$ with probability $\frac{1}{2}$ and lottery $d_3$ with probability $\frac{1}{2}$, show that no utility function $u$ can exist for the group when two members have the utility functions given above.

# 6

# Bayesian statistics for decision analysis

## 6.1 INTRODUCTION

We illustrated in Section 4.5 how probability statements about uncertain events can be made more precise by 'credence decomposition'. Credence decomposition breaks probability statements down into small tractable units which the client can handle. These probability statements are then manipulated by the analyst to derive probabilities that are important in the solution of her problem.

One of the most important credence decompositions, which we have not yet discussed, I shall call the basic statistical model (BSM). Although not all statistical models are BSM, nearly all of the common ones are. So they form the cornerstone to all statistical inferences.

## 6.2 THE BASIC STATISTICAL MODEL

A random vector $\mathbf{X}_1 = (X_1, X_2, \ldots, X_n)$ will have been observed before you need to take a decision. A random vector $\mathbf{X}_2 = (X_{n+1}, \ldots, X_{n+k})$ of random variables influences the outcome of any decision you choose to take and is not independent of $\mathbf{X}_1$. Hence you require to specify your distribution of $\mathbf{X}_2 | (\mathbf{X}_1 = \mathbf{x}_1)$.

Now, of course, you could try to specify the distribution of $\mathbf{X}_2 | (\mathbf{X}_1 = \mathbf{x}_1)$ directly, but this is going to be very difficult if the dependence between $\mathbf{X}_1$ and $\mathbf{X}_2$ is at all complicated. In the *basic statistical model* we introduce a vector of random variables $\boldsymbol{\theta} = (\theta_1, \theta_2, \ldots, \theta_m)$, $\boldsymbol{\theta} \in \Theta \subseteq \mathbb{R}^m$, whose task it is to represent all the variation *shared* between $(\mathbf{X}_1, \mathbf{X}_2) = (X_1, X_2, \ldots, X_{n+k})$. In particular, if you were to know $\boldsymbol{\theta}$ then the distribution of $\mathbf{X}_2$ would no longer depend on $\mathbf{X}_1$.

Mathematically the BSM states that conditional on $\boldsymbol{\theta}, X_1, X_2, \ldots, X_{n+k}$ are independent (written $\coprod_{i=1}^{n+k} X_i | \boldsymbol{\theta}$).

In terms of an influence diagram the relationship between $\mathbf{X}_1$ and $\mathbf{X}_2$ is represented in Fig. 6.1 when $n = 4$ and $k = 1$.

As can be seen from the diagram, even when there are a small number of observable random variables the description using the explanatory variables $\boldsymbol{\theta}$ is much simpler than the description without its construction.

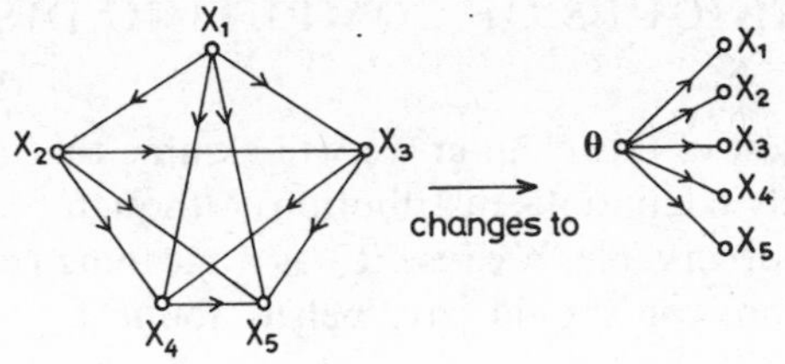


**Figure 6.1** Influence diagram of five random variables conditionally independent of $\boldsymbol{\theta}$

## 6.3 EXAMPLES OF BASIC STATISTICAL MODELS

1. The random vector $\boldsymbol{\theta}$ represents a true *transmitted* signal. $\mathbf{X}_1$ denotes various readings, with error, received from the transmitter. $\mathbf{X}_2$ represents the future readings that will be received.
2. The random vector $\boldsymbol{\theta}$ represents the level of impurities in a chemical vat. $\mathbf{X}_1$ denotes the levels of impurities measured in that vat to date, $\mathbf{X}_2$ represents the level of impurities in future samples.
3. The random vector $\boldsymbol{\theta}$ represents the 'true' ability of a student. $\mathbf{X}_1$ denotes his performance in past tasks/exams and $\mathbf{X}_2$ his performance in future tasks/exams.
4. The random variable $\theta$ denotes the proportion of people in a population with a certain disease. $\mathbf{X}_1$ denotes the observed proportions in samples of the population and $\mathbf{X}_2$ denotes the proportion of people with the disease we expect to meet in the future.
5. The random variable $\theta$ denotes the mean yield of milk from an unspecified cow in a specified herd. $\mathbf{X}_1$ denotes the observed yields of individual cows in the herd at various days and $\mathbf{X}_2$ denotes the yield of various cows in future days.
6. The random variable $\theta$ denotes the true level of demand per week for a product a client manufactures. $\mathbf{X}_1$ denotes his past sales and $\mathbf{X}_2$ his future sales of that product.
7. Computer users log into a mainframe at a rate $\theta$ at a certain time. $\mathbf{X}_1$ denotes the past record of users log-on times and $\mathbf{X}_2$ the future log-on times on the computer.

In decision analysis problems you are typically interested in the distribution of $\mathbf{X}_2|(\mathbf{X}_1 = \mathbf{x}_1)$ and sometimes also in the distribution of $\boldsymbol{\theta}|(\mathbf{X}_1 = \mathbf{x}_1)$. Note that sometimes the explanatory variable $\boldsymbol{\theta}$ could in practice never be observed (see Examples 3, 4, 5, 6 above). The vector $\boldsymbol{\theta}$ is then just a useful tool which helps you think about your problem and so construct an appropriate joint distribution between $\mathbf{X}_1$ and $\mathbf{X}_2$.

## 6.4 THE TECHNIQUES OF COMPUTING DISTRIBUTIONS GIVEN $\mathbf{X}_1 = \mathbf{x}_1$

Throughout the rest of this chapter we will assume that the random vector $\boldsymbol{\theta}$ has an absolutely continuous distribution. Although the random vector $\boldsymbol{\theta}$ may not even be observable, because it has a meaning (see the examples of the previous section) you should have beliefs about it.

*Definition*

Your *prior density* (represented by $f(\boldsymbol{\theta})$) represents your beliefs about the vector $\boldsymbol{\theta}$ before you observe either $\mathbf{X}_1$ or $\mathbf{X}_2$.

The prior density can of course be very vague about the values $\boldsymbol{\theta}$ can take. This is often the case when dealing with straight problems of inference (e.g. see Lindley, 1980). On the other hand, you may know from its definition that the random vector $\boldsymbol{\theta}$ must, with high probability, take values in a small range. For example in Example 1 of Section 6.3 you may know that the signal you are observing obeys certain characteristics. Examples of how you might specify a prior density are given after equation (6.9) and in the example just before Section 6.8.

Your next task is to specify the distribution of $\mathbf{X}_1$ given $\boldsymbol{\theta}$. This is usually much easier than specifying the distribution of $\mathbf{X}_1$ because by conditioning on $\boldsymbol{\theta}$ you have modelled all systematic variation across your observations. If you are given $\boldsymbol{\theta}$, all that is usually left to model is the 'random' fluctuations from observation to observation.

*Definition*

Let $p(\mathbf{x}_1|\boldsymbol{\theta})$ denote the probability mass function or probability density function of $\mathbf{X}_1$ given $\boldsymbol{\theta}$ of discrete or absolutely continuous $\mathbf{X}_1$ respectively.

The *likelihood* of $\boldsymbol{\theta}|\mathbf{x}_1$ is any positive function $l(\boldsymbol{\theta}|\mathbf{x}_1)$ of $\boldsymbol{\theta}$ proportional to $p(\mathbf{x}|\boldsymbol{\theta})$ (thought of as a function of $\boldsymbol{\theta}$). By the Bayes rule,

$$f(\boldsymbol{\theta}|\mathbf{x}_1) = (p(\mathbf{x}_1))^{-1} f(\boldsymbol{\theta}) p(\mathbf{x}_1|\boldsymbol{\theta}) \tag{6.1}$$

where $f(\boldsymbol{\theta}|\mathbf{x}_1)$ is the density of $\boldsymbol{\theta}$ given $\mathbf{x}_1$ (called the *posterior density* of $\boldsymbol{\theta}$), $p(\mathbf{x}_1)$ is the probability mass function or density function of $\mathbf{X}_1$, and $f(\boldsymbol{\theta})$ is the prior density of $\boldsymbol{\theta}$ defined above.

Hence your beliefs about your explanatory variables $\boldsymbol{\theta}$ posterior to observing $\mathbf{x}_1$ are given probabilistically by the density $f(\boldsymbol{\theta}|\mathbf{x}_1)$ by the formula (6.1).

Thinking of equation (6.1) as an expression in $\boldsymbol{\theta}$ it can be somewhat simplifed into the equation

$$f(\boldsymbol{\theta}|\mathbf{x}_1) \propto f(\boldsymbol{\theta}) l(\boldsymbol{\theta}|\mathbf{x}_1) \tag{6.2}$$

where $l(\boldsymbol{\theta}|\mathbf{x}_1)$ is any likelihood of $\boldsymbol{\theta}$ given $\mathbf{x}_1$ or (provided no function on either side of the equation is zero)

$$\log f(\boldsymbol{\theta}|\mathbf{x}_1) = \log f(\boldsymbol{\theta}) + \lambda(\boldsymbol{\theta}|\mathbf{x}_1) + h(\mathbf{x}_1) \tag{6.3}$$

where $h(\mathbf{x}_1)$ is a function in $\mathbf{x}_1$ only which ensures $f(\boldsymbol{\theta}|\mathbf{x}_1)$ integrates to unity and $\lambda = \log l$.

Now under the BSM, $\coprod_{i=1}^{n} X_i|\boldsymbol{\theta}$, so in particular

$$p(\mathbf{x}_1|\boldsymbol{\theta}) = \prod_{i=1}^{n} p_i(x_i|\boldsymbol{\theta})$$

where $p_i$ is the mass function or density of discrete or absolutely continuous $X_i$ respectively, i $\leqslant i \leqslant n$, so

$$l(\boldsymbol{\theta}|\mathbf{x}_1) = \prod_{i=1}^{n} l_i(\boldsymbol{\theta}|x_i)$$

where $l_i$ is any function of $\boldsymbol{\theta}$ proportional to $p_i(x_i|\boldsymbol{\theta})$, and (provided $l_i$ does not take the value zero at any observation $x_i$),

$$\lambda(\boldsymbol{\theta}|\mathbf{x}_1) = \sum_{i=1}^{n} \lambda_i(\boldsymbol{\theta}|x_i) \tag{6.4}$$

where $\lambda_i = \log l_i$.

Substituting (6.4) into (6.3) gives us the formula

$$\log f(\boldsymbol{\theta}|x_1) = \log f(\boldsymbol{\theta}) + \sum_{i=1}^{n} \lambda_i(\boldsymbol{\theta}|x_i) + h(\mathbf{x}_1) \tag{6.5}$$

where $\lambda_i(\boldsymbol{\theta}|\mathbf{x}_i) - \log p_i(x_i|\boldsymbol{\theta})$ is constant in $\boldsymbol{\theta}$ and $h(\mathbf{x}_1)$ is a function in $\mathbf{x}_1$ which ensures $f(\boldsymbol{\theta}|\mathbf{x}_1)$ integrates to unity.

I shall illustrate the use of formula (6.5) in the next section. Firstly, however, I will derive the formula that will give us the distribution of $\mathbf{X}_2|\mathbf{X}_1$ given the BSM.

Under the BSM, because $\coprod_{i=1}^{2} \mathbf{X}_i|\boldsymbol{\theta}$,

$$p(\mathbf{x}_2|\boldsymbol{\theta}, \mathbf{x}_1) = p(\mathbf{x}_2|\boldsymbol{\theta}) \tag{6.6}$$

where $p$ is the mass function or density of discrete or continuous random variables $\mathbf{X}_2|\boldsymbol{\theta}$ respectively. By extending the conversation over the absolutely continuous $\boldsymbol{\theta}|\mathbf{x}_1$, the joint mass function/density function of $\mathbf{X}_2|\mathbf{X}_1$, $p(\mathbf{x}_2|\mathbf{x}_1)$ is given by

$$p(\mathbf{x}_2|\mathbf{x}_1) = \int_{\theta \in \Theta} p(\mathbf{x}_2|\boldsymbol{\theta}, \mathbf{x}_1) f(\boldsymbol{\theta}|\mathbf{x}_1)\, \mathrm{d}\boldsymbol{\theta} \tag{6.7}$$

where $f(\boldsymbol{\theta}|\mathbf{x}_1)$ is the posterior density of $\boldsymbol{\theta}$ given $\mathbf{x}_1$. So by equation (6.6),

$$p(\mathbf{x}_2|\mathbf{x}_1) = \int_{\theta \in \Theta} p(\mathbf{x}_2|\boldsymbol{\theta}) f(\boldsymbol{\theta}|\mathbf{x}_1)\, \mathrm{d}\boldsymbol{\theta} \tag{6.8}$$

Again, the distribution of $\mathbf{X}_2|\boldsymbol{\theta}$ is usually easier to specify than the distribution of $\mathbf{X}_2$ since by conditioning on $\boldsymbol{\theta}$ we should have removed all systematic variation between the components of $\mathbf{x}_2$.

## 6.5 THE NORMAL BASIC STATISTICAL MODEL

Suppose $\coprod_{i=1}^{n} X_i|\theta$ where each $X_i$ has a normal distribution with mean $\theta$ and variance $V$. For Example 2 of Section 6.2, $x_1, x_2, \ldots, x_n$ might be measurements of the level of a given impurity from $n$ samples (with replacement) from a well-mixed vat of chemical. The random variable $\theta$ represents the level of impurity averaged over the whole vat and $V$ measures the variability across the vat of our sample reading. We will assume $V$ is known.

Since $X_1, \ldots, X_n$ are normally distributed,

$$p(\mathbf{x}_1|\theta) = \prod_{i=1}^{n} p_i(x_i|\theta)$$

where

$$p_i(x_i|\theta) = \frac{1}{\sqrt{2\pi V}} \exp\{-\tfrac{1}{2}V^{-1}(x_i - \theta)^2\}$$

One choice of log likelihood $\lambda_i$ of the $i$th observation is

$$\lambda_i = -\tfrac{1}{2}V^{-1}(x_i - \theta)^2 \tag{6.9}$$

(note $e^{\lambda_i} \propto p_i$ here, as required by the definition of $\lambda_i$).

Now, for the sake of simplicity, assume that your beliefs about $\theta$ can be adequately expressed by saying that you believe that $\theta$ has a normal distribution with mean $m$ and variance $W$. Having as yet not observed $\mathbf{X}_2$ you expect that $\theta$ takes a value $m$ but are obviously uncertain about this to the extent fixed by $W$, viz., for example,

$$p\{|\theta - m| \leqslant 2\sqrt{W}\} \simeq 0.95$$

Since $\theta$ is normally distributed it will follow that it has a prior density of the form

$$f(\theta) = \frac{1}{\sqrt{2\pi W}} \exp\{-\tfrac{1}{2}W^{-1}(\theta - m)^2\} \tag{6.10}$$

### 6.5.1 Prior to posterior analysis

You now simply use equation (6.5) to obtain the posterior density $f(\theta|\mathbf{x}_1)$. Thus using equations (6.9) and (6.10),

$$\log f(\theta|\mathbf{x}_1) = -\tfrac{1}{2}\log(2\pi W) - \tfrac{1}{2}W^{-1}(\theta - m)^2 + \sum_{i=1}^{n} -\tfrac{1}{2}V^{-1}(\theta - x_i)^2 + h(\mathbf{x}_1)$$

where $h(\mathbf{x}_1)$ is some function of $\mathbf{x}_1$ only

$$= -\tfrac{1}{2}[W^{-1}(\theta - m)^2 + \sum_{i=1}^{n} V^{-1}(\theta - x_i)^2] + h'(\mathbf{x}_1)$$

where $h'(\mathbf{x}_1) = h(\mathbf{x}_1) - \tfrac{1}{2}\log(2\pi W)$.

Now clearly the terms in $\theta$ in this expression are at most quadratic. It is easily checked that the above equation can be rewritten in the form

$$\log f(\theta|\mathbf{x}_1) = -\tfrac{1}{2}[W_1^{-1}(\theta - m_1)^2] + h''(\mathbf{x}_1) \tag{6.11}$$

where $W_1 = (W^{-1} + nV^{-1})^{-1}$

$$m_1 = (W_1)^{-1}(W^{-1}m + nV^{-1}\bar{x}), \quad \text{where} \quad \bar{x} = n^{-1}\sum_{i=1}^{n} x_i$$

$$h''(\mathbf{x}_1) = -\tfrac{1}{2}\left[(W + n^{-1}V)^{-1}(\bar{x} - m_1)^2 + nV^{-1}\sum_{i=1}^{n}(x_i - \bar{x})^2\right] + h'(\mathbf{x})$$

So, thought of as a function of $\theta$, exponentiating equation (6.11) we obtain

$$f(\theta|\mathbf{x}_1) \propto \exp\{-\tfrac{1}{2}[W_1^{-1}(\theta - m_1)^2]\} \tag{6.12}$$

where $W_1$ and $m_1$ are given above. So the density of $\theta|\mathbf{x}_1$ is proportional to a normal density with mean $m_1$ and variance $W_1$. But since a density must integrate to one this implies that $f(\theta|\mathbf{x}_1)$ must be a normal density with mean $m_1$ and variance $V_1$. Note that the observations $x_1$ only influence your beliefs about $\theta$ through our expectation $m_1$ of $\theta$ and then only through the sample mean $\bar{x}$ of our observations.

### 6.5.2 A predictive distribution

Suppose now that a customer will sample your vat of chemicals and that the amount of chemical he buys will be dependent on the measurement $X_{n+1}$ of impurity he observes. The BSM allows you to assume that $\coprod_{i=1}^{n+1} X_i|\theta$. Assume also that $X_{n+1}$ is normally distributed like the other samples with mean $\theta$ and variance $V$. Using equation (6.8) you now have that

$$p(x_{n+1}|\mathbf{x}_1) = \int_{-\infty}^{\infty} p(x_{n+1}|\theta) f(\theta|\mathbf{x}_1)\,d\theta$$

$$= (2\pi)^{-1}V^{-1/2}W_1^{-1/2}\int_{-\infty}^{\infty} \exp\{-\tfrac{1}{2}\{V^{-1}(\theta - x_{n+1})^2 + W^{-1}(\theta - m_1)^2\}\}\,d\theta \tag{6.13}$$

by equations (6.9) and (6.12), where $m_1$ and $W_1$ are defined above. Again, with a little algebra it is easy to check that

$$V^{-1}(\theta - x_{n+1})^2 + W_1^{-1}(\theta - m_1)^2 = W_2^{-1}(\theta - m_2)^2 + (V + W)^{-1}(x_{n+1} - m_1)^2$$

where

$$m_2 = W_2^{-1}(V^{-1}x_{n+1} + W_1^{-1}m_1) \tag{6.14}$$

$$W_2 = (V^{-1} + W_1^{-1})^{-1} \tag{6.15}$$

Hence equation (6.13) becomes

$$p(x_{n+1}|\mathbf{x}_1) = h(x_{n+1}|\mathbf{x}_1)\int_{-\infty}^{\infty} \exp -\tfrac{1}{2}\{W_2^{-1}(\theta - m_2)^2\}\,\mathrm{d}\theta$$

where

$$h(x_{n+1}|\mathbf{x}_1) = (2\pi)^{-1}V^{-1/2}W^{-1/2}\exp\{-\tfrac{1}{2}(V+W)^{-1}(x_{n+1}-m_1)^2\} \tag{6.16}$$

But the integrand above is proportional to a normal density with mean $m_2$ and variance $W_2$. Its integral must therefore be $\sqrt{2\pi W_2}$. Hence

$$p(x_{n+1}|\mathbf{x}_1) = h(x_{n+1})\sqrt{2\pi W_2}$$

which using equation (6.16) can be rewritten

$$= \{2\pi(V+W_1)\}^{-1/2}\exp\{-\tfrac{1}{2}(V+W_1)^{-1}(x_{n+1}-m_1)^2\}$$

Hence $X_{n+1}|\mathbf{x}_1$ has a normal density with mean $m_1$ and variance $V + W_1$ where $m_1$ and $W_1$ are given in equation (6.11).

### 6.5.3 Exercises on the normal BSM

1. Show that your posterior mean $m_1$ of the impurity in your vat is a weighted average of your prior mean $m$ and sample mean $\bar{x}$ of the form

   $$m_1 = A\bar{x} + (1-A)m \qquad \text{where } A = nV^{-1}\{nV^{-1} + W^{-1}\}^{-1}$$

   Show, in particular, that as $n \to \infty$ for fixed $V$ and $W > 0$, your posterior mean $m_1 \to \bar{x}$.
2. Show that $W_1 < W$, i.e. your variance on $\theta$ always decreases after you observe $\mathbf{x}_1$.
3. Show that the variance of $X_{n+1}|\mathbf{X}_1 = \mathbf{x}_1$ tends to the variance $V$ of $X_{n+1}|\theta$ as $n \to \infty$. Interpret this result in words.

## 6.6 CONVENIENT RELATIONSHIPS IN PRIOR TO POSTERIOR ANALYSIS

In the last section we had the rather convenient property that the posterior distribution of $\theta$ came from the same family, the normal family, as the prior distribution of $\theta$. This provokes the following definition.

*Definition*

A family of prior densities $F = \{f(\boldsymbol{\theta}|\boldsymbol{\alpha}): \alpha \in A \subseteq \mathbb{R}^s\}$ is said to be *closed under sampling* for likelihood $l(\boldsymbol{\theta}|\mathbf{x}_1)$ if for any prior density $f(\boldsymbol{\theta})$ in $F$ and any data $\mathbf{x}_1$ whose likelihood is greater than zero, the corresponding posterior density $f(\boldsymbol{\theta}|\mathbf{x}_1)$ is also in $F$.

**Advantage of using families closed under sampling** It is very easy to see how your data influences your client's beliefs about the explanatory variable $\boldsymbol{\theta}$. For example, in the normal example above you only need compare the means and variances of the prior and posterior densities of $\boldsymbol{\theta}$ to obtain a complete picture of how your client's distribution of $\boldsymbol{\theta}$ changes in the light of the data $\mathbf{x}_1$.

**Disadvantage** Simple prior families closed under sampling may be too restrictive to represent a client's beliefs accurately. For example, in Section 6.5 it may not be legitimate to assume your client's beliefs are *a priori* normally distributed. Computer programs are now available to do the grind of the calculation of posterior densities when priors are not closed under sampling (see e.g. Naylor and Shaw, 1985, 1986).

*Examples of densities closed under sampling*

(a) Normal densities are closed under sampling for normal likelihoods (see Section 6.5).
(b) The family of all prior densities which assigns a probability zero to any set not intersecting the set $A$ is closed under sampling for any continuous likelihood.

The following theorem lists a useful family of prior densities for some commonly occurring likelihoods.

*Theorem 6.1*

Suppose a log likelihood $\lambda(\boldsymbol{\theta}|\mathbf{x}_1)$ satisfies the following equation for all possible sample points $\mathbf{x}_1$ in a set $X$:

$$\lambda(\boldsymbol{\theta}|\mathbf{x}_1) = \begin{cases} \boldsymbol{\tau}(\mathbf{x}_1)\boldsymbol{\psi}(\boldsymbol{\theta}) + k_1(\mathbf{x}_1) & \mathbf{x}_1 \in X \\ -\infty & \text{otherwise} \end{cases} \tag{6.17}$$

where $\boldsymbol{\tau}(\mathbf{x}_1)$ is the same $1 \times m$ real vector function of $\mathbf{x}_1$ only, $\boldsymbol{\psi}(\boldsymbol{\theta})$ is some $m \times 1$ real vector function of $\boldsymbol{\theta}$ only and $k_1(\mathbf{x}_1)$ is not a function of $\boldsymbol{\theta}$.

Let $F$ be a family of densities indexed by $\boldsymbol{\alpha} \in A$ given by

$$f(\boldsymbol{\theta}|\boldsymbol{\alpha}) = \exp\{\boldsymbol{\alpha}\cdot\boldsymbol{\psi}(\boldsymbol{\theta}) + k_2(\boldsymbol{\alpha})\} \qquad \boldsymbol{\alpha} \in A \tag{6.18}$$

where $\boldsymbol{\alpha}$ is some $1 \times m$ real vector not depending on $\boldsymbol{\theta}$.

Then provided $\boldsymbol{\alpha} + \boldsymbol{\tau}(\mathbf{x}_1) \in A$ for all $\mathbf{x}_1 \in X$ (i.e. for all values of $\mathbf{x}_1$ whose likelihood is not zero), then $F$ is closed under sampling for $\lambda(\boldsymbol{\theta}|\mathbf{x}_1)$. Furthermore, the parameters $\boldsymbol{\alpha}_1$ of the posterior density of $\boldsymbol{\theta}$ are linked to the parameter $\boldsymbol{\alpha}$ of the prior density by the equation

$$\boldsymbol{\alpha}_1 = \boldsymbol{\alpha} + \tau(\mathbf{x}_1) \tag{6.19}$$

**Proof** The result follows simply by applying equation (6.3). Thus

$$\begin{aligned}\log f(\boldsymbol{\theta}|\mathbf{x}_1, \boldsymbol{\alpha}) &= \log f(\boldsymbol{\theta}|\boldsymbol{\alpha}) + \lambda(\boldsymbol{\theta}|\mathbf{x}_1) + h(\mathbf{x}_1, \boldsymbol{\alpha}) \\ &= \boldsymbol{\alpha}\boldsymbol{\psi}(\boldsymbol{\theta}) + k_2(\boldsymbol{\alpha}) + \tau(\mathbf{x}_1)\boldsymbol{\psi}(\boldsymbol{\theta}) + k_1(\mathbf{x}_1) + h(\mathbf{x}_1, \boldsymbol{\alpha})\end{aligned}$$

by equations (6.17) and (6.18)

$$= \boldsymbol{\alpha}_1 \boldsymbol{\psi}(\boldsymbol{\theta}) + h_1(\mathbf{x}_1, \boldsymbol{\alpha})$$

where $\boldsymbol{\alpha}_1$ is defined above and $h_1(\mathbf{x}_1, \boldsymbol{\alpha}) = h + k_1 + k_2$

Hence $f(\boldsymbol{\theta}|\mathbf{x}_1, \boldsymbol{\alpha}_1) = \exp\{\boldsymbol{\alpha}_1 \boldsymbol{\psi}(\boldsymbol{\theta}) + h_2(\boldsymbol{\alpha}_1)\}$

where $h_2(\boldsymbol{\alpha}_1)$ ensures $f(\boldsymbol{\theta}|\mathbf{x}_1 \boldsymbol{\alpha}_1)$ integrates to 1.

A family of densities satisfying equation (6.18) is sometimes called a *conjugate family* to a log likelihood $\lambda$.

## 6.7 SOME EXAMPLES OF THE USE OF THEOREM 6.1

### 6.7.1 The normal BSM

$\coprod_{i=1}^{n} x_i|\theta$ normal, mean $\theta$, variance $V$ known. We have already done the prior to posterior analysis for this model in Section 6.5. Here I show that you get the same result somewhat quicker using the theorem. A log likelihood $\lambda(\theta|\mathbf{x}_1)$ can be written in the form

$$\lambda(\theta|\mathbf{x}_1) = -\tfrac{1}{2}V^{-1}(\theta - x_i)^2 = \boldsymbol{\tau}\boldsymbol{\psi}(\theta + k_1(\mathbf{x}_1)$$

where

$$\boldsymbol{\tau} = (-\tfrac{1}{2}nV^{-1}, nV^{-1}\bar{x}),\ \boldsymbol{\psi} = (\theta^2, \theta)' \quad \text{and} \quad k_1(\mathbf{x}_1) = -\tfrac{1}{2}V^{-1}\sum_{i=1}^{n} x_i^2$$

The normal prior density can also be written in the appropriate form. Thus

$$f(\theta|\boldsymbol{\alpha}) = (2\pi W)^{-1/2}\exp\{-\tfrac{1}{2}W^{-1}(\theta - m)^2\} = \exp\{\boldsymbol{\alpha}\boldsymbol{\psi}(\theta) + k_2(\boldsymbol{\alpha})\}$$

where

$$\boldsymbol{\alpha} = (-\tfrac{1}{2}W^{-1}, W^{-1}m), k_2(\boldsymbol{\alpha}) = -\tfrac{1}{2}\log(2\pi W) - \tfrac{1}{2}W^{-1}m^2. \tag{6.20}$$

From Theorem 6.1 therefore

$$f(\boldsymbol{\theta}|\boldsymbol{\alpha}, \mathbf{x}_1) = f(\theta|\boldsymbol{\alpha}_1)$$

where $\boldsymbol{\alpha}_1 = \boldsymbol{\alpha} + \boldsymbol{\tau} = (-\tfrac{1}{2}(W^{-1} + nV^{-1}), W^{-1}m + nV^{-1}\bar{x})$.

Using the relationship between $\boldsymbol{\alpha}$ and $(W, m)$ given in equation (6.20) the posterior variance $W_1$ and mean $m_1$ of $\theta$ are thus given by

$$-\tfrac{1}{2}W_1^{-1} = -\tfrac{1}{2}(W^{-1} + nV^{-1})$$

implying

$$W_1 = (W^{-1} + nV^{-1})^{-1} \tag{6.21}$$

and

$$W_1^{-1} m_1 = W^{-1} m + nV^{-1} \bar{x}$$

implying

$$m_1 = W_1(W^{-1} m + nV^{-1} \bar{x}) \tag{6.22}$$

These equations of course agree with your original prior to posterior analysis.

### 6.7.2 The beta-binomial BSM

Suppose $\coprod_{i=1}^{n} X_i | \theta$ and the density of each random variable $X_i$ is given by

$$p_i(x_i | \theta) = \binom{N}{x_i} \theta^{x_i}(1-\theta)^{N-x_i} \qquad x_i = 1, 2, \ldots, n, 0 < \theta < 1$$

$$\lambda_i(\theta | x_i) = \log p_i(x_i | \theta) = \log \binom{N}{x_i} + x_i \log \theta + (N - x_i) \log (1-\theta)$$

and

$$\lambda(\theta | \mathbf{x}_1) = \sum_{i=1}^{n} \lambda_i(\theta | x_i) = \boldsymbol{\tau} \boldsymbol{\psi}(\theta) + k_1(\mathbf{x}_1)$$

where

$$\boldsymbol{\tau} = (n\bar{x}, n(N - \bar{x})), \boldsymbol{\psi}(\theta) = (\log \theta, \log (1-\theta))' \text{ and } k_1(\mathbf{x}_1) = \sum_{i=1}^{n} \log \binom{N}{x_i} \tag{6.23}$$

where $\bar{x} = n^{-1} \sum_{i=1}^{n} x_i$

Assume that the prior density of $\theta$ has a beta density written

$$f(\theta | \alpha) = \frac{\Gamma(\alpha + \beta)}{\Gamma(\alpha)\Gamma(\beta)} \theta^{\alpha-1}(1-\theta)^{\beta-1} \qquad \begin{matrix} \alpha, \beta > 0 \\ 0 < \theta < 1 \end{matrix} \tag{6.24}$$

where $\Gamma(\alpha) = \int_0^\infty u^{\alpha-1} e^{-u} \, du$, $\alpha > 0$.

This can be rewritten in the form

$$f(\theta | \boldsymbol{\alpha}) = \exp \{\boldsymbol{\alpha} \boldsymbol{\psi}(\theta) + k_2(\boldsymbol{\alpha})\}$$

where

$$\boldsymbol{\alpha} = (\alpha - 1, \beta - 1), k_2(\boldsymbol{\alpha}) = \log \Gamma(\alpha + \beta) - \log \Gamma(\alpha) - \log \Gamma(\beta)$$

From Theorem 6.1 therefore the posterior density of $\theta$ is also beta. Let

$$f(\theta | \boldsymbol{\alpha}, \mathbf{x}_1) = f(\boldsymbol{\theta} | \boldsymbol{\alpha}_1)$$

then $\boldsymbol{\alpha}_1 = \boldsymbol{\alpha} + \boldsymbol{\tau} = (\alpha + n\bar{x} - 1, \beta + n(N - \bar{x}) - 1)$.

*Exercises on the beta density*

1. The mean $m$ of the beta density given in equation (6.24) is $\alpha(\alpha+\beta)^{-1}$. Show that the mean $m_1$ of $\theta$ posterior to observing $\mathbf{x}_1$ is given by the equation

$$m_1 = (1-A)m + AN^{-1}\bar{x}$$

where $A = nN[\alpha+\beta+nN]^{-1}$.

Show that $m_1 \to N^{-1}x$ as $nN \to \infty$.

2.* Use Chebychev's inequality to show that for all $\varepsilon > 0$

$$p\{|m_1 - \theta| \geqslant \varepsilon\} \to 0 \quad \text{as } nN \to \infty$$

*Note*: the variance of the beta density given in equation (6.24) is $\alpha\beta(\alpha+\beta)^{-2}(\alpha+\beta+1)^{-1}$.

### 6.7.3 The Poisson/exponential BSM

Let $X_1 = (X_1, \ldots, X_m, Y_1, \ldots, Y_n)$ and $\coprod (X_1, \ldots, X_m, Y_1, \ldots, Y_n)|\theta$ where $X_i$, $1 \leqslant i \leqslant m$, has an exponential density given by

$$p_i(x|\theta) = \theta e^{-x\theta} \qquad \theta > 0, x > 0,$$

and $Y_i$, $1 \leqslant i \leqslant m$, has a Poisson mass function

$$p_i(y_i|\theta) = \frac{\theta^y}{y!} e^{-\theta} \qquad \theta > 0, y = 0, 1, 2, \ldots$$

Then it is easily checked that a log likelihood for $\theta$ given $\mathbf{x}_1$ is given by

$$\lambda(\theta|\mathbf{x}_1) = \boldsymbol{\tau}\boldsymbol{\psi}(\theta)$$

where

$$\boldsymbol{\tau} = (m + n\bar{y}, n + m\bar{x}) \qquad \boldsymbol{\psi}'(\theta) = (\log\theta, -\theta) \qquad \bar{y} = n^{-1}\sum_{i=1}^{n} y_i$$

$$x = m^{-1}\sum_{i=1}^{m} x_i$$

If you have a gamma prior density on $\theta$ of the form

$$f(\theta|\alpha) = \frac{\beta^\alpha}{\Gamma(\alpha)}\theta^{\alpha-1}e^{-\beta\theta} \qquad \theta > 0, \alpha, \beta > 0. \tag{6.25}$$

this can be written in conjugate form for the theorem. Thus

$$f(\theta|\boldsymbol{\alpha}) = \exp\{\boldsymbol{\alpha}\boldsymbol{\psi}(\theta) + k_2(\boldsymbol{\alpha})\}$$

where

$$\boldsymbol{\alpha} = (\alpha - 1, \beta) \qquad \boldsymbol{\psi}'(\theta) = (\log\theta, -\theta) \qquad k_2(\boldsymbol{\alpha}) = \alpha\log\beta - \log\Gamma(\alpha) \tag{6.26}$$

From Theorem 6.1 therefore the posterior density for $\theta$ is also gamma with new parameters given by

$$\boldsymbol{\alpha}_1 = \boldsymbol{\alpha} + \boldsymbol{\tau} = (\alpha + m + n\bar{y} - 1, \beta + n + m\bar{x}) \tag{6.27}$$

In particular, your posterior mean $m_1$ of $\theta$ is given by

$$m_1 = \frac{\alpha + m + n\bar{y}}{\beta + n + m\bar{x}}$$

since $\alpha/\beta$ is the mean of the gamma density given in equation (6.25).

*A predictive distribution*

Now suppose $n = 0$ so that we only observe the exponential random variables $X_1, \ldots, X_m$ and we are interested in the distribution of $X_{m+1}|X_1, \ldots, X_m$. Using equation (6.8) we see that the predictive density $f_{m+1}(x|\mathbf{x}_1)$ is given by

$$f_{m+1}(x|\mathbf{x}_1) = \int_0^\infty \theta e^{-\theta x} f(\theta|\boldsymbol{\alpha}_1, \mathbf{x}_1)\,\mathrm{d}\theta \qquad x > 0$$

$$= \int_0^\infty \theta e^{-\theta x} \frac{\beta_1^{\alpha_1}}{\Gamma(\alpha_1)} \theta^{\alpha_1 - 1} e^{-\beta_1 \theta}\,\mathrm{d}\theta$$

where by equation (6.27)

$$\alpha_1 = \alpha + m \quad \text{and} \quad \beta_1 = \beta + m\bar{x} \tag{6.28}$$

Rearranging,

$$f_{m+1}(x|\mathbf{x}_1) = \frac{\beta_1^{\alpha_1}}{\Gamma(\alpha_1)} \int_0^\infty \theta^{\alpha_1} e^{-(\beta_1 + x)\theta}\,\mathrm{d}\theta \tag{6.29}$$

Now, from the definition of the gamma density in equation (6.25), since this density must integrate to one we have the equality

$$\int_0^\infty \theta^{\alpha_1 - 1} e^{-\beta\theta}\,\mathrm{d}\theta = \frac{\Gamma(\alpha)}{\beta^\alpha}$$

It follows that

$$\int_0^\infty \theta^{\alpha_1} \mathrm{e}^{-(\beta_1 + x)\theta}\,\mathrm{d}\theta = \frac{\Gamma(\alpha_1 + 1)}{(\beta_1 + x)^{\alpha_1 + 1}} \tag{6.30}$$

It is easily shown that $\Gamma(\alpha_1 + 1) = \alpha_1 \Gamma(\alpha_1)$, so combining equations (6.29) and (6.30) we have that

$$f_{m+1}(x|\mathbf{x}_1) = \frac{\alpha_1 \beta_1^{\alpha_1}}{(\beta_1 + x)^{\alpha_1 + 1}} \qquad x > 0 \tag{6.31}$$

where $\alpha_1$ and $\beta_1$ are given in equation (6.27). This density is called a *Pareto density*.

**Table 6.1** Some common one-parameter prior to posterior analyses

| Name: Observations' density (mass function)/prior density | A set of $n$ independent observations $x$ with density/mass function proportional to | Conjugate prior with density proportional to | Prior mean $m$ | Prior variance $W$ | Posterior mean $m_1$ | Posterior variance $W_1$ |
|---|---|---|---|---|---|---|
| Normal/normal | $\exp\{-\frac{1}{2}V^{-1}(x-\theta)^2\}$ | $\exp\{-\frac{1}{2}W^{-1}(\theta-m)^2\}$ | $m$ | $W$ | $A\bar{x}+(1-A)m$ $(A=nW_1/V)$ | $(W^{-1}+nV^{-1})^{-1}$ |
| Poisson/gamma | $\frac{\theta^x}{x!}\exp\{-\theta\}$ $x=0,1,2,\ldots$ | $\theta^{\alpha-1}\exp\{-\beta\theta\}$ $\theta>0$ | $\alpha/\beta$ | $\alpha/\beta^2$ | $A\bar{x}+(1-A)m$ $(A=n(\beta+n)^{-1})$ | $\frac{m_1}{\beta+n}$ |
| Exponential/gamma | $\theta\exp(-x\theta)$ $(x>0)$ | $\theta^{\alpha-1}\exp\{-\beta\theta\}$ $\theta>0$ | $\alpha/\beta$ | $\alpha/\beta^2$ | $[A\bar{x}+(1-A)m^{-1}]^{-1}$ $(A=n(\alpha+n)^{-1})$ | $\frac{m_1}{\beta+n\bar{x}}$ |
| Binomial/beta | $\binom{N}{x}\theta^x(1-\theta)^{N-x}$ $x=0,1,\ldots,N$ | $\theta^{\alpha-1}(1-\theta)^{\beta-1}$ $0\leqslant\theta\leqslant 1$ | $\frac{\alpha}{\alpha+\beta}$ | $\frac{m(1-m)}{\alpha+\beta+1}$ | $AN^{-1}\bar{x}+(1-A)m$ $A=\frac{nN}{\alpha+\beta+nN}$ | $\frac{m_1(1-m_1)}{\alpha+\beta+nN+1}$ |
| Uniform/Pareto | $\begin{cases}\theta^{-1} & 0\leqslant x\leqslant\theta\\ 0 & \text{otherwise}\end{cases}$ | $\begin{cases}\theta^{-\alpha-1} & \theta>\beta\\ 0 & \text{otherwise}\end{cases}$ $(\alpha>2,\beta>0)$ | $\frac{\alpha\beta}{\alpha-1}$ | $\frac{m^2}{\alpha(\alpha-2)}$ | $\frac{(\alpha+n)\beta_1}{(\alpha+n-1)}$ $\beta_1=\max\{\beta,x_1,\ldots,x_n\}$ | $\frac{m_1^2}{(\alpha+n)(\alpha+n-1)}$ |

*Exercise*

Find the mean and variance of $X_{m+1}|\mathbf{x}_1$.

*Example*

You have just bought a machine which will be able to do a job for a client. The job will need 4 hours to run and you can only afford to run the job once. Unfortunately the machine you have bought is very delicate and may break down some time after you have started the job. The breakdown rate per hour $\theta$ varies depending on where it is used but you are told by the manufacturer that when used in a variety of locations the machine broke down an average of 10 times every 100 hours. The variance $V$ of $\theta$ is reported as 0.01.

You have done your own quality control on the machine you bought. It was run five times until it broke down and the durations of each run were (in hours) 3.2, 12.7, 20.6, 7.9, 10.2.

Find the probability that the next breakdown will occur before the job is finished, assuming that:

(a) the times $X_i$, $1 \leqslant i \leqslant 5$, to breakdown are independent and exponential with rate $\theta$;
(b) it is reasonable to assume that $\theta$ has, *a priori*, a gamma distribution.

**Answer** The mean and variance of the gamma density given in equation (6.25) are respectively $\alpha/\beta$ and $\alpha/\beta^2$. Since the mean and variance of $\theta$ are given to be 0.1 and 0.01 respectively in this problem, your prior density for $\theta$ has $\alpha = 0.01/(0.1)^2 = 1$ and $\beta = 0.1/0.01 = 10$.

Using equation (6.27) we can see that posterior to our quality control $\theta$ has a gamma density with new parameters $\alpha_1$ and $\beta_1$ given by

$$\alpha_1 = 5, \beta_1 = 10 + 3.2 + 12.7 + 20.6 + 7.9 + 10.2 = 64.6$$

The density of the time $X_6$ before the next breakdown is given by equation (6.31). By integrating this density we find that

$$P(X_6 \leqslant 4) = 1 - \frac{\beta_1^{\alpha_1}}{(\beta_1 + 4)^{\alpha_1}} = 0.26$$

is our required probability.

A summary of some of the more common prior to posterior analyses is given in Table 6.1. Further examples of conjugate analyses can be found in DeGroot (1970) and Aitchison and Dunsmore (1975).

## 6.8* FINDING THE POSTERIOR JOINT DENSITY OF A RANDOM VECTOR OF EXPLANATORY VARIABLES

In principle Theorem 6.1 applies equally to a random vector $\boldsymbol{\theta}$ as it does to a single random variable $\theta$. The algebra just gets a little more messy. We here give two examples of important prior to posterior analyses. The results of these examples rather than the technical details are what is important here, so we will skip through the necessary algebra quickly. More detail can be found in DeGroot (1970), Raiffa and Schlaifer (1961), Press (1972) and Aitchison and Dunsmore (1975). (An excellent simple account of Bayes linear models is given in Lindley, 1980.) These references also contain many more general examples of prior to posterior analyses beyond the scope of this book.

### 6.8.1 The normal, variance unknown BSM

In Section 6.5 we assumed that the variance of $X_1, \ldots, X_n$ was known to be $V$. This is not usually a realistic assumption and $V$ will need to be estimated. So $V$ now becomes one of our explanatory variables.

Suppose $\coprod_{i=1}^{n} X_i|(\theta_1, \theta_2)$ are normally distributed random variables with mean $\theta_1$ and variance $\theta_2$. The density of $X_i$ is then

$$P_i(x|\theta) = \frac{1}{\sqrt{2\pi\theta_2}} \exp\{-\tfrac{1}{2}\theta_2^{-1}(\theta_1 - x)^2\} \qquad \theta_2 > 0$$

It follows that one log likelihood $\lambda(\boldsymbol{\theta}|\mathbf{x}_1)$ for this independent sample is given by

$$\lambda(\boldsymbol{\theta}, \mathbf{x}_1) = \boldsymbol{\tau}\boldsymbol{\psi}(\boldsymbol{\theta}) \tag{6.32}$$

where

$$\boldsymbol{\tau} = \left( -n/2, \frac{1}{2}\sum_{i=1}^{n} x_i^2, \sum_{i=1}^{n} x_i, -n/2 \right) \tag{6.33}$$

and

$$\boldsymbol{\psi}'(\boldsymbol{\theta}) = (\log\theta_2, -\theta_2^{-1}, \theta_2^{-1}\theta_1, \theta_2^{-1}\theta_1^2) \tag{6.34}$$

Suppose our prior density on $\boldsymbol{\theta}$ is in the convenient form

$$f(\boldsymbol{\theta}) = f(\theta_1, \theta_2) = f(\theta_1|\theta_2) f(\theta_2)$$

where

$$f(\theta_1|\theta_2) = (2\pi\tau^{-1}\theta_2)^{-1/2} \exp\{-\tfrac{1}{2}\theta_2^{-1}\tau(\theta_1 - m)^2\}, \qquad \theta_2 > 0, \tau > 0$$

$$f(\theta_2) = \frac{\beta^\alpha}{\Gamma(\alpha)}\theta_2^{-(\alpha+1)} \exp\{-\beta\theta_2^{-1}\} \qquad \alpha, \beta > 0 \tag{6.35}$$

Thus conditional on the variance $\theta_2$ the mean of $X_1, \ldots, X_n$ has a normal distribution with expectation $m$ and variance $\tau^{-1}\theta_2$.

Marginally, *a priori*, we believe the inverse of the variance $\theta_2^{-1}$ has gamma density with parameters $(\alpha, \beta)$. This peculiar looking joint density happens to be conjugate to $\lambda$ because it can be written in the form

$$f(\boldsymbol{\theta}) = \exp\{\boldsymbol{\alpha}\boldsymbol{\psi}(\boldsymbol{\theta}) + k_2(\boldsymbol{\alpha})\} \tag{6.36}$$

where

$$\boldsymbol{\alpha} = (-\alpha - 3/2, \beta + \tfrac{1}{2}\tau m^2, \tau m, -\tfrac{1}{2}\tau)$$

and $k(\alpha)$ is some complicated function constant in $\boldsymbol{\theta}$ and $\boldsymbol{\psi}(\boldsymbol{\theta})$ is given in equation (6.34).

It follows from our theorem that the density of $\boldsymbol{\theta}$ given $\mathbf{x}_1$ has the same form as that given by equation (6.35) where $\boldsymbol{\alpha}$ is replaced by $\boldsymbol{\alpha}_1$, where

$$\boldsymbol{\alpha}_1 = \boldsymbol{\alpha} + \boldsymbol{\tau} = \Bigg( -\tfrac{1}{2}\alpha - 3/2 - n/2, \beta + \tfrac{1}{2}\tau m^2 + \frac{1}{2}\sum_{i=1}^{n} x_i^2, \tau m + \sum_{i=1}^{n} x_i, -\tfrac{1}{2}(n+\tau) \Bigg)$$

After some messy algebra, we obtain the posterior values $(\alpha', \beta', m', \tau')$ of $(\alpha, \beta, m, \tau)$ in the original forms of density given in equation (6.35) as

$$\alpha' = \alpha + \tfrac{1}{2}n \tag{6.37}$$

$$\beta' = \beta + \frac{1}{2}\sum_{i=1}^{n}(x_i - \bar{x})^2 + \frac{\tau n(\bar{x} - m)^2}{2(\tau + n)} \tag{6.38}$$

$$m' = (\tau')^{-1}\left(\tau m + \sum_{i=1}^{n} x_i\right) \tag{6.39}$$

$$\tau' = \tau + n \tag{6.40}$$

Posterior to observing $\mathbf{x}_1$ the distribution of $\theta_1 | \theta_2$ and $\theta_2^{-1}$ are normal and gamma respectively, with new parameters given above.

**Comments** To unravel the meaning of this algebra, note the following point. We can rewrite equation (6.39) to give us that

$$m' = Am + (1 - A)\bar{x}$$

where $A = \tau(\tau + n)^{-1}$ and $\bar{x} = n^{-1}\sum_{i=1}^{n} x_i$. So the mean of $\theta_1 | \theta_2$ given $\mathbf{x}_1$ is a weighted average of the prior mean of $\theta_1 | \theta_2$ and the sample mean $\bar{x}$ of the observations. As $n \to \infty$ so $m' \to \bar{x}$, but for moderate values of $n$ we adjust this posterior mean towards our prior guess about the mean of $\theta_1 | \theta_2$.

Similarly, the prior mean $V$ of the variance $\theta_2$ can be shown to be $\beta/\alpha + 1$. The expected value or 'best guess' of the variance $V'$ of $\boldsymbol{\theta}_2 | \mathbf{x}_1$ is related to $V$ by the formula

$$V' = AV + (1 - A)S^2 + \frac{\tau n(\bar{x} - m)^2}{(\tau + n)(2\alpha + n + 2)} \tag{6.41}$$

where

$$A = \frac{\alpha + 1}{\alpha_1 + \tfrac{1}{2}n + 1} \quad \text{and} \quad S^2 = n^{-1}\sum_{i=1}^{n}(x_i - \bar{x})^2.$$

The last term in equation (6.41) tends quickly to zero as $n \to \infty$. So our best estimate of $\operatorname{Var}(\theta_2 | \mathbf{x}_1)$ is a weighted average of our prior estimate of $\operatorname{Var}(\theta_2)$

and the sample variance together with a positive adjustment factor which is large only if your prior guess $m$ of $\theta_1 | \theta_2$ is very different from the sample mean.

## 6.9* THE MULTIVARIATE NORMAL EXCHANGEABLE MODEL

Here I give one important illustration of a prior to posterior analysis when there are many explanatory random variables of interest to us.

*Definitions*

A *multivariate normal density* is any density $f(\boldsymbol{\theta})$ on the random vector $\boldsymbol{\theta} = (\theta_1, \theta_2, \ldots, \theta_n)$ taking the form

$$f(\boldsymbol{\theta}) \propto \exp -\tfrac{1}{2}\{Q(\boldsymbol{\theta})\} \tag{6.42}$$

where $Q(\boldsymbol{\theta})$ is a quadratic function in $\boldsymbol{\theta}$ of a form which allows $f(\boldsymbol{\theta})$ to be integrable.

Random variables $\theta_1, \theta_2, \ldots, \theta_n$ are said to be *exchangeable* if the permutation of any two indices in the components of $\boldsymbol{\theta}$ leaves the distribution of $\theta_1, \ldots, \theta_n$ unchanged.

It is appropriate to use an exchangeable prior distribution on $\boldsymbol{\theta}$ when, for example, your client has prior information on the group of attributes $\boldsymbol{\theta}$, but no extra information on the individual members comprising that group. Suppose he has bought several machines of the same type to do several similar jobs. His information from the manufacturer will be about the set of machines as a group, not about each individual machine he has bought. His knowledge will therefore be exchangeable across the machines. Any probability statement about one machine is the same as the corresponding probability statement on another.

In another instance an external examiner might have prior information about the quality of a class of students sitting an exam but no extra information about the individuals comprising that class. Again a prior distribution over measures of ability of the $i$th member of the class could sensibly be assumed to be exchangeable.

It is not difficult to check that multivariate normal random variables are exchangeable if and only if their joint density $f(\boldsymbol{\theta})$ is given by equation (6.42) where $Q(\boldsymbol{\theta})$ has the special form

$$Q(\boldsymbol{\theta}) = (na)^{-1}\left\{\sum_{i=1}^{n} \theta_i^2 + n^{-1}b \sum_{i=1}^{n}\sum_{j=1}^{n} \theta_i\theta_j - 2(1+b)m \sum_{i=1}^{n} \theta_i\right\} \tag{6.43}$$

where the conditions $a > 0$ and $-1 < b < n(n-1)^{-1}$ ensure that $Q(\boldsymbol{\theta})$ is of a form to allow $f(\boldsymbol{\theta})$ to be integrable.

Exchangeable random variables all share the same mean $m$ and variance $W$. In our notation using multivariate normal theory it can be shown that the variance $W$ of $\theta_i$ is given by

$$W = \frac{a[n + (n-1)b]}{1 + b} \tag{6.44}$$

If the normal random variables $\theta_1, \ldots, \theta_n$ were independent then $b = 0$ in our notation. On the other hand, not all exchangeable random variables are independent of each other. In particular, when the density has $Q(\boldsymbol{\theta})$ defined as above the correlation between any two random variables is given by the equation

$$\rho = -b(n + (n-1)b)^{-1} \tag{6.45}$$

So the closer $b$ is to $-1$, the more highly correlated are the components $\theta_i$.

Let us now suppose we take a sample $\coprod_{i=1}^{n} X_i | \boldsymbol{\theta}$ so that $X_i$ has a normal density $p_i(x_i|\theta_i)$ given by

$$P_i(x_i|\theta_i) = \frac{1}{\sqrt{2\pi V}} \exp\{-\tfrac{1}{2}V^{-1}(\theta_i - x_i)^2\} \qquad V > 0 \tag{6.46}$$

In our first example we would have performed a single efficiency test on each of our machines. The result of this test has value $x_i$ whose distribution had mean $\theta_i$, the 'true efficiency of the machine', and variance $V$. In our second classroom example, the external examiner will have seen the scores $x_i$ of the $i$th member of the class which he might assume was approximately normally distributed with mean $\theta_i$, his true ability, and variance $V$.

From equation (6.46) we can deduce that the log likelihood, defined in equation (6.4) is of the form

$$\lambda(\boldsymbol{\theta}|\mathbf{x}) = -\tfrac{1}{2}R(\boldsymbol{\theta})$$

where

$$R(\boldsymbol{\theta}) = V^{-1} \sum_{i=1}^{n} \theta_i^2 - 2 \sum_{i=1}^{n} (V^{-1}x_i)\theta_i$$

From equation (6.43) we now have that the joint density $f(\boldsymbol{\theta}|\mathbf{x}_1)$ of $\boldsymbol{\theta}$ after observing $\mathbf{x}_1$ is of the form

$$f(\boldsymbol{\theta}|\mathbf{x}) \propto \exp -\tfrac{1}{2}\{S(\boldsymbol{\theta})\}$$

where

$$\begin{aligned} S(\boldsymbol{\theta}) &= O(\boldsymbol{\theta}) + R(\boldsymbol{\theta}) \\ &= ((na)^{-1} + V^{-1}) \sum_{i=1}^{n} \theta_i^2 + n^{-2}ba^{-1} \sum_{i=1}^{n} \sum_{j=1}^{n} \theta_i\theta_j \\ &\qquad - 2 \sum_{i=1}^{n} [(na)^{-1}(1+b)m + Vx_i]\theta_i \end{aligned}$$

We now try to interpret this density of $\boldsymbol{\theta}|\mathbf{x}_1$. Since $f(\boldsymbol{\theta}|\mathbf{x}_1)$ happens to be symmetric in all its components and is unimodal, it is not difficult to show that its vector of mean values $\mathbf{m}_1$ is just its joint mode. This joint mode can be located by differentiating $S(\boldsymbol{\theta})$ partially with respect to each of its components in turn and then solving the $n$ equation in $\boldsymbol{\theta}$ obtained by putting these derivatives to zero. Well,

$$\frac{\partial S(\boldsymbol{\theta})}{\partial \theta_i} = 2[(na)^{-1} + V^{-1}]\theta_i + 2n^{-2}ba^{-1}\sum_{i=1}^{n}\theta_i - 2[(na)^{-1}(1+b)m + V^{-1}x_i] \qquad 1 \leqslant i \leqslant n$$

giving us the $n$ equations

$$[(na)^{-1} + V^{-1}]\theta_i + n^{-2}ba^{-1}\sum_{i=1}^{n}\theta_i = (na)^{-1}(1+b)m + V^{-1}x_i \qquad 1 \leqslant i \leqslant n \tag{6.47}$$

We first solve for $\sum_{i=1}^{n}\theta_i$ by adding up these $n$ equations. We can then solve for the individual modes of $\theta_i$ by substituting this value back into equation (6.46). Adding equation (6.46) gives

$$[(na)^{-1} + V^{-1}]\sum_{i=1}^{n}\theta_i + n^{-1}ba^{-1}\sum_{i=1}^{n}\theta_i = a^{-1}(1+b)m + nV^{-1}\bar{x}$$

where $\bar{x} = n^{-1}\sum_{i=1}^{n}x_i$. Thus our expectation of $m_1 = n^{-1}(\sum_{i=1}^{n}\theta_i)$ is

$$m_1 = Bm + (1-B)\bar{x} \qquad \text{where } B = \frac{(na)^{-1}(1+b)}{(na)^{-1}(1+b) + V^{-1}} \tag{6.48}$$

Substituting this into equation (6.47) thus gives the posterior mean $\mathbf{m}_1$ of $\boldsymbol{\theta}|\mathbf{x}_1$ as having $i$th term $m(i)$ given by

$$m(i) = Ax_i + (1-A)[m - b(m_1 - m)] \tag{6.49}$$

where $A = na(na+V)^{-1}$ and $m_1$ is given above.

If we write equations (6.48) and (6.49) in terms of the prior variance $W$ $1 - A[m - b(m_1 - m)]$ and prior correlation $\beta$ across $\boldsymbol{\theta}$, after a little algebra it can be shown by using equations (6.44) and (6.45) that $b = -n(n-1+\rho^{-1})^{-1}$

$$B = \frac{V}{W(1+(n-1)\rho) + V}$$

and

$$A = \frac{W(1-\rho)}{W(1-\rho) + V}$$

In particular, as the number of observations $n \to \infty$ if $\rho \neq 0$ and $W$ are fixed then the posterior mean of $\theta_i$, $m(i) \to Ax_i + (1-A)\bar{x}$, a weighted average of the $i$th observation and the sample mean of the observations. So, in our first example, our beliefs about the reliability quotient $\theta_i$ of the $i$th machine depends on not just the result of the $i$th test, but also the results of the tests of all the other machines. Similarly, in the second example, the external examiner should assess each student as a weighted average of his mark and the average mark of the class. The amount you adjust towards the sample mean $\bar{x}$ depends on the value you assign to the prior correlation $\rho$ between the abilities of any two students. Further detailed discussion of the theory and applications of exchangeable priors can be found in Lindley and Smith (1972).

## EXERCISES

6.1 Elementary particles are emitted independently from a nuclear source. If $X_1$ denotes the time before the first emission, in minutes, and $X_i$ denotes the time between the emission of the $(i-1)$th and $i$th particle, $i=2,3,4,\ldots$, we know that the density $f_1(x|\theta)$ of each $X_i$ is given by

$$f_1(x|\theta)=\begin{cases}\theta e^{-\theta x} & x>0, \theta>0\\ 0 & \text{otherwise}\end{cases}$$

When you first obtain the data you find that you have only been given the number $Y_j$ of emissions in the time interval $(j-1, j]$. You know, however, that from the above $Y_j$, $j=1,2,\ldots$, are independent with mass function $f_2(y|\theta)$, where

$$f_2(y|\theta)=\frac{\theta^y}{y!}e^{-\theta} \qquad y=0,1,2,\ldots, \qquad \theta>0.$$

You take $m$ observations and your last observation arrives exactly on the $n$th minute. Your prior distribution for $\theta$ was a gamma distribution $G(\alpha^*, \beta^*)$ with density $\pi(\theta)$, where

$$\pi(\theta)\propto\begin{cases}\theta^{\alpha^*-1}\exp(-\beta^*\theta) & \theta>0, \alpha^*, \beta^*>0\\ 0 & \text{otherwise}\end{cases}$$

Find your posterior distribution for

(i) using $y_1,\ldots,y_n$
(ii) using $x_1,\ldots,x_m$

and show that the two analyses give identical inferences on $\theta$. If $\alpha^*>1$ express the posterior mode of $\theta$ as a weighted average of $y$ and the prior mode (where the weight $\rho$ depends only on $\beta^*/n$).

6.2 (i) Show that if a family of prior densities with members $f(\boldsymbol{\theta}|\boldsymbol{\alpha})$, $\boldsymbol{\alpha}\in A$, is closed under sampling for a likelihood $l(\boldsymbol{\theta}|\mathbf{x}_1)$, then the family of prior densities whose members are

$$f(\boldsymbol{\theta}|\boldsymbol{\alpha},\lambda_1,\ldots,\lambda_n)=\sum_{i=1}^{k}\lambda_i f(\boldsymbol{\theta}|\boldsymbol{\alpha}_i) \qquad \alpha_i\in A, \qquad 1\leqslant i\leqslant n, \qquad 0<\lambda_i,$$

$$\sum_{i=1}^{n}\lambda_i=1$$

is also closed under sampling for a likelihood $l(\boldsymbol{\theta}|\mathbf{x}_1)$.

(ii) For integer $(\alpha_1, \alpha_2)$ the beta $(\alpha_1, \alpha_2)$ density $f(\theta|\boldsymbol{\alpha})$ is given by

$$f(\theta|\alpha) = \begin{cases} \dfrac{(\alpha_1 + \alpha_2 - 1)!}{(\alpha_1 - 1)!(\alpha_2 - 1)!}\theta^{\alpha_1 - 1}(1 - \theta)^{\alpha_2 - 1} & 0 < \theta < 1 \\ 0 & \text{otherwise} \end{cases}$$

Your prior density that the probability $\theta$ of a coin coming up heads on any toss is given by

$$f(\theta) = \tfrac{1}{2} f(\theta|(1, 1)) + \tfrac{1}{2} f(\theta|(\alpha, \alpha))$$

where $\alpha$ is a large integer. Find your posterior density $f(\theta|x)$ of $\theta$ given you observe $x$ heads in $n$ independent tosses of the coin and show it can be written in the form

$$f(\theta|x) = \lambda_1 f(\theta|\boldsymbol{\alpha}_1) + \lambda_2 f(\theta|\boldsymbol{\alpha}_2) \qquad \lambda_1 + \lambda_2 = 1, \qquad \lambda_1, \lambda_2 > 0$$

Show in particular that if $\alpha \geqslant 2$ and $x = 0$, as the number $n$ of tosses increases, $\lambda_1$ tends to unity. How does the evolution of the shape of $f(\theta|x)$ differ as $n \to \infty$, from the evolution of the posterior density of $\theta$ had we used a 'conjugate' beta prior given in Section 6.7, Example 2?

6.3 $X$ is a normal random variable with mean $\theta$ and unit variance. Your prior density $f(\theta)$ for $\theta$ is given by

$$f(\theta) = \tfrac{1}{2}[g(\theta| -\mu, 1) + g(\theta|\mu, 1)]$$

where $g(\theta|m, W)$ is the normal density with mean $m$ and variance $W$. Find $f(\theta|\mathbf{x})$ and show that the posterior variance $V$ of $\theta$ satisfies

$$V \leqslant \tfrac{1}{2}[1 + \tfrac{1}{2}\mu^2] \qquad \text{with equality if and only if } x = 0$$

If you had used a normal prior density with the same mean and variance as $f(\theta)$, how would your posterior density of $\theta$ vary from the one above?

6.4 $\coprod_{i=1}^{n+1} X_i|\theta$ each with a rectangular density $f(x|\theta)$ given by

$$f(x|\theta) = \begin{cases} \theta^{-1} & 0 < x < \theta \\ 0 & \text{otherwise} \end{cases}$$

Your prior density $f(\theta)$ of $\theta$ is given by

$$f(\theta) = \begin{cases} \alpha\theta_0^{\alpha}\theta^{-(\alpha+1)} & \theta > \theta_0, \quad \alpha > 0, \quad \theta_0 > 0 \\ 0 & \text{otherwise} \end{cases}$$

Find the posterior density of $\theta$ given $x_1, \ldots, x_n$ and the predictive density of $x_{n+1}$ given $x_1, x_2, \ldots, x_n$. Sketch this predictive density.

6.5* Random variables $\coprod_{i=1}^{n} X_i|\boldsymbol{\theta}$, $\boldsymbol{\theta} = (\theta_1, \theta_2)$ each have a normal density with mean $\theta_1$ and variance $\theta_2$. Instead of using a conjugate analysis on $\boldsymbol{\theta}$

as given in Section 6.8 assume that $\theta_1$ and $\theta_2$ are *a priori* independent with respective densities

$$f_1(\theta_1) = (2\pi\omega)^{-1/2} \exp\{-\tfrac{1}{2}\omega^{-1}(\theta_1 - m)^2\}$$

$$f_2(\theta_2) = \frac{\beta^\alpha}{\Gamma(\alpha)} \theta_2^{-(\alpha+1)} \exp\{-\beta\theta_2^{-1}\}$$

Show that the marginal posterior density $f(\theta_1|\mathbf{x})$ of $\theta_1$ satisfies

$$f(\theta_1|\mathbf{x}) = q(\theta_1) r(\theta_1)$$

where $q(\theta_1)$ is a normal density and $r(\theta_1)$ is a $t$-density. Prove that there are values of $\mu$, $W$, $\mathbf{x}$, $\alpha$ and $\beta$ which cause this density to have two modes.

6.6 $\coprod_{i=1}^n X_i|\boldsymbol{\theta}$, $\boldsymbol{\theta} = (\theta_1, \theta_2, \ldots, \theta_n)$ where $X_i$ has a normal distribution with mean $\theta_i$ and variance $V$, $1 \leqslant i \leqslant n$. *A priori* you believe that

$$\begin{aligned} &\theta_0 = 0 \\ &\theta_i = \theta_{i-1} + \varepsilon_i \qquad 1 \leqslant i \leqslant n \end{aligned}$$

where $\varepsilon_1, \ldots, \varepsilon_n \coprod$ each with zero mean and variance $W$. (This type of prior distribution is often used in Bayesian forecasting; see Harrison and Stevens, 1976). Show that the posterior density $f(\boldsymbol{\theta}|\mathbf{x})$ is given by

$$f(\boldsymbol{\theta}|\mathbf{x}_1) \propto \exp\left\{-\tfrac{1}{2}W^{-1}\sum_{i=1}^n(\theta_i - \theta_{i-1})^2 + V^{-1}\sum_{i=1}^n(x_i - \theta_i)^2\right\}$$

Since the function in the exponent is quadratic in $\boldsymbol{\theta}$, $f(\boldsymbol{\theta}|\mathbf{x})$ is a normal density. Give an iterative solution for the $i$th component $m_i$ of the posterior mode vector $(m_1, \ldots, m_n)$ of this joint density and calculate $m_1, m_2, m_3, m_4$ when $x_1 = 1$, $x_2 = 10$, $x_3 = 1$, $x_4 = -1$, $V = 1$ and $W = \tfrac{1}{4}$. Note that this prior's estimate of the mean of $X_i$ smooths over *all* the observations.

6.7 Calculate the posterior mode vector in the normal exchangeable model given in Section 6.9 when each random variable $X_i$, $1 \leqslant i \leqslant n$, is independently replicated $K$ times.

6.8* (*Return to our drilling example of Chapter* 2) We are now in a position to make the example in Chapter 2 more realistic. Instead of his original assumption, suppose your client believes before experimentation that the log volume of oil $V(A)$, $V(B)$ from fields $A$ and $B$ respectively have approximate normal distributions with respective (mean, variance) parameters (5, 1), (25, 25). He can choose to investigate either field (but not both) by taking a reading $X$. When $X$ is taken from field $A$ it has mean $V(A)$ and variance 1. A net profit of \$10 million is expected per unit volume. The cost of an investigation is \$6 million and the cost of the option to drill \$31 million as before.

Using your influence diagram of this problem together with the normal conjugate analysis given in Section 6.5 calculate the EMV strategy as a function (if necessary) of the reading $X$.

6.9 You have just brought three heavy-duty sewing machines of an identical type. You are interested in the fastest speed, $\theta_i$, you can drive the $i$th machine, $i = 1, 2, 3$, over an average piece of cloth of a given thickness without breaking a typical needle on the machine. From the information given to you by the manufacturer about the sewing machines, you can deduce that the joint density $f(\boldsymbol{\theta})$ between $\boldsymbol{\theta} = (\theta_1, \theta_2, \theta_3)$ is exchangeable and given by

$$f(\boldsymbol{\theta}) = (2\pi)^{-3/2}(\sqrt{3})^{-1} \exp\left\{-\frac{1}{2}\left[\sum_{i=1}^{3} (\theta_i - \bar{\theta})^2 + (\bar{\theta} - \mu)^2\right]\right\}$$

where $\bar{\theta} = (\theta_1 + \theta_2 + \theta_3)/3$.

Find the posterior means of $\theta_1$, $\theta_2$, $\theta_3$, in the two cases:

(a) You test the speed $\theta_i$ of each machine once on a piece of cloth, obtaining the fastest speed $y_i, 1 < i < 3$ before your needle breaks. You assume $y_1, y_2$, and $y_3$ are independent and normally distributed with $y_i$ having mean $\theta_i$ and variance 3, $1 < i < 3$.

(b) You test the first machine three times on three different combinations of pieces of cloth and needle. The needle breaks at speeds of $z_1, z_2, z_3$ respectively. You assume $z_1, z_2, z_3$ are all independent normal random variables with mean $\theta_1$ and variance 3.

Show that the two posterior means $E(\bar{\theta}|y)$ and $E(\bar{\theta}|z)$ can be written in the form

$$E(\bar{\theta}|\mathbf{y}) = \rho\mu + (1-\rho)\bar{y}, \qquad 0 < \rho < 1, \bar{y} = (y_1 + y_2 + y_3)/3$$
$$E(\bar{\theta}|\mathbf{z}) = \sigma\mu + (1-\sigma)\bar{z}, \qquad 0 < \sigma < 1, \bar{z} = (z_1 + z_2 + z_3)/3$$

Find $\rho$ and $\sigma$ and show that $\rho < \sigma$. Why is this the case?

# 7

# Bayes estimation

## 7.1 INTRODUCTION

Statisticians are often asked to give a point estimate for a random variable $\theta$, that is a best guess of the value of $\theta$. To do this in a Bayesian framework you need to first specify your losses consequent on making a decision $d$ when various values of $\theta$ pertain. The best point estimate under the EMV algorithm is then decision $d^*$, called the *Bayes decision*, which minimizes the expected loss – the expectation being taken across the distribution of $\theta$.

## 7.2 THE QUADRATIC LOSS FUNCTION

A convenient loss function whose point estimate and its associated expected loss are linked to well-known summary quantities of a distribution is the quadratic loss function $Q(\theta, d) = (d - \theta)^2$.

*Theorem 7.1*

If the variance, $\text{Var}(\theta)$, of $\theta$ exists then the Bayes estimate $d^*$ is the mean, $\mu$ of $\theta$. The associated expected loss on taking the Bayes decision is $\text{Var}(\theta)$.

**Proof** The expected loss $\bar{Q}$ is given by

$$\begin{aligned}
\bar{Q}(d) &= E(\theta - d)^2 \qquad \text{where the expectation is taken across } \theta \\
&= E[(\theta - \mu) + (\mu - d)]^2 \\
&= E(\theta - \mu)^2 + (\mu - d)^2 + 2(\mu - d)E(\theta - \mu) \\
&= \text{Var}(\theta) + (\mu - d)^2 + 2(\mu - d)(E(\theta) - \mu) \\
&= \text{Var}(\theta) + (\mu - d)^2 \qquad \text{since } \mu = E(\theta)
\end{aligned}$$

So clearly $\bar{Q}(d) \geqslant \text{Var}(\theta)$ with equality if and only if $d = \mu$. Thus the Bayes decision $d^* = \mu$ and its associated expected loss is $\text{Var}(\theta)$.

Typically when estimating the value of the random variable $\theta$, the distribution of $\theta$ will be posterior to observing a vector of random variables $\mathbf{x}_1$. When this is the case, $\mu$ and $\text{Var}(\theta)$ will be functions of $\mathbf{x}$. Table 6.1 lists Bayes decisions (means) and their consequent expected losses (variances) under quadratic loss for a variety of distributions of data and conjugate linear prior densities.

Although it is simple in most cases to find a Bayes decision when using the quadratic loss function, it is difficult to justify the use of this loss function in most applications. However, in many simple decision problems it is reasonable to assume that the appropriate loss function is approximately piecewise linear in $\theta$ and $d$. The simplest piecewise linear loss function is the absolute loss function.

## 7.3 THE ABSOLUTE LOSS FUNCTION

The asymmetric absolute loss function $A(\theta, d|a_1, a_2)$ is defined by

$$A(\theta, d|a_1, a_2) = \begin{cases} a_1(\theta - d) & \theta > d \\ a_2(d - \theta) & \theta \leqslant d \end{cases} \quad \text{where } a_1, a_2 > 0 \tag{7.1}$$

Note that $A$ is greater than or equal to zero for all values of $d$ and $\theta$ and that no loss is incurred only when $d = \theta$, i.e. when you estimate $\theta$ correctly.

*Example 7.1*

Your client needs to order stock. Every item he does not sell will need to be stored at a cost of $\$a_2$. Every item he might be able to sell but doesn't because he is out of stock will lose him $\$a_1$ net profit. The variable $\theta$ represents the number of items he will be able to sell whose distribution function is $F$, and $d$ represents his decision of the number of items he orders. If you want to follow the EMV algorithm then you should choose $d$ to minimize your expected absolute loss given in equation (7.1).

The following theorem is important in helping to find Bayes decisions under absolute loss.

Call a point $d^*$ an $a_1/(a_1 + a_2)$ percentile of $\theta$ if

$$P(\theta \leqslant d^*) \geqslant \frac{a_1}{a_1 + a_2} \tag{7.2}$$

$$P(\theta \geqslant d^*) \geqslant \frac{a_2}{a_1 + a_2}$$

*Theorem 7.2*

Under absolute loss given in equation (7.1) a decision $d$ is a Bayes decision if and only if it is an $a_1/(a_1 + a_2)$ percentile of $\theta$, provided that $E|\theta| = M$ where $M$ is finite.

**Note** If $a_1 = a_2$ and the loss function is symmetric then this theorem implies that the Bayes decision is a median of $\theta$.

The proof of the theorem is a little technical and so has been relegated to Appendix 3.

The result of this theorem is very useful. Consider the following example.

*Example 7.2*

The sales in the next quarter of a certain product are known to be approximately normally distributed with mean $\mu$ and variance $\sigma^2$. Let

$$X = \sigma^{-1}(\theta - \mu) \tag{7.3}$$

Then $X$ is normally distributed with zero mean and unit variance. Let $C_\alpha$ be defined by the equation

$$P(X \leqslant C_\alpha) = \alpha \qquad 0 < \alpha < 1$$

The values of $C_\alpha$ for various values of $\alpha$ are tabulated in all good sets of statistical tables. From equation (7.3) the $\alpha = a_1/(a_1 + a_2)$ percentile $d^*$ of $\theta$ is related to $C_\alpha$ by

$$\sigma^{-1}(d^* - \mu) = C_\alpha$$

i.e.

$$d^* = \mu + C_\alpha \sigma$$

Return to our stock ordering problem of Example 7.1. Suppose $a_1 = 3$ and $a_2 = 1$, making $\alpha = 3/4$. This would mean that the cost of storing an item of stock is a third of the cost of losing the sale of that item. Since $C_{3/4} = 0.773$, with the sales $\theta$ distributed normally as above, the Bayes decision is to choose to order

$$d^* = \mu + 0.773\sigma$$

Notice that your client should choose to order more than he expects to sell because losses arising from storage costs are less than losses arising from lost sales. As $\sigma \to 0$ and he becomes more and more certain of his expected sales, so obviously the optimal order $d^* \to \mu$.

## 7.4 BOUNDED LOSS FUNCTIONS AND WHY WE SHOULD USE THEM

A *bounded loss function* is a loss function which is bounded above and below for all possible values of $\theta$ and $d$. Both quadratic loss functions and absolute loss functions are not bounded loss functions since in each of these cases $L(\theta, d) \to \infty$ as $|\theta - d| \to \infty$.

One reason why loss functions that are not bounded cause problems is that their expected losses $L$ might be infinite for all values of $d$. For example, if $\theta$ has a distribution $F(\theta)$ given by

$$F(\theta) = \begin{cases} 0 & \theta \leqslant 1 \\ 1 - \theta^{-1} & \theta > 1 \end{cases}$$

then it is easily checked that $E(\theta) = E|\theta|$ is infinite and that the expected loss with respect to either quadratic or an absolute loss function is always infinite regardless of the decision. So we cannot reasonably advise any decision in this case. In fact we have to include many technical and unintuitive conditions on axiomatic systems on utilities and losses we met in Chapter 3 before we can justify using unbounded loss functions.

Furthermore, Bayes decisions arising from unbounded loss functions can change enormously when the distribution of our random variable changes infinitesimally (see for example Kadane and Chuang, 1978, or Smith, 1978). An example of this instability is given in Exercise 7.1. So we have to be absolutely precise about our probability statements, which are in practice difficult to measure, before we can be confident of providing a sensible estimate. In any case, in real-life situations usually it will not be possible to lose an infinite amount. For example, if our loss is monetary we will have a fixed amount of capital we can lose before going bankrupt. Henceforth we will assume that all our loss functions are bounded. This will avoid all the problems mentioned above.

Since Bayes decisions are invariant under increasing linear transformations of our utility or loss function (see Section 3.4), without loss of generality we shall, for the rest of the chapter assume that $L(\theta, d)$ is bounded below by 0 and above by 1.

## 7.5 THE STEP LOSS FUNCTION

The *step loss function* $S_b(\theta - d)$ is defined by

$$S_b(\theta - d) = \begin{cases} 0 & |\theta - d| \leqslant b \\ 1 & \text{otherwise.} \end{cases}$$

Step loss is a very simple bounded loss function. If you misestimate $\theta$ by an amount less than $b$ you lose nothing. If you don't you lose all you can and the estimate is considered inadequate. A Bayes decision associated with $S_b$ is quite easily linked to the density of a continuous random variable $\theta$, as the following theorem shows.

*Theorem 7.3*

If a random variable $\theta$ has a density $f$ which is continuous on the real line then any Bayes decision $d^*$ under loss $S_b$ will satisfy

$$f(d^* + b) = f(d^* - b) \tag{7.4}$$

**Proof** Let the expected loss associated with $S_b$ be denoted by $S_b(d)$. Then

$$\bar{S}_b(d) = \int_{-\infty}^{\infty} S_b(\theta - d) f(\theta)\, d\theta$$

$$= \int_{-\infty}^{\infty} S_b(s) f(s+d)\,\mathrm{d}s \qquad \text{where } s = \theta - d$$

$$= \int_{\infty}^{-b} f(s+d)\,\mathrm{d}s + \int_{b}^{\infty} f(s+d)\,\mathrm{d}s \qquad \text{from the definition of } S_b$$

$$= F(d-b) + 1 - F(d+b) \tag{7.5}$$

where $F$ is the distribution function of $\theta$. The Bayes decision must be a stationary point of this function. In particular, if $F$ is differentiable everywhere as is implied from the condition of the theorem, the stationary points of $\bar{S}_b(d)$ will be given by

$$\frac{\mathrm{d}}{\mathrm{d}x}\bar{S}_b(x) = 0$$

which from equation (7.5) gives us that

$$f(d^* - b) = f(d^* + b)$$

for a Bayes decision $d^*$, as required.

**Notes** Firstly note that equation (7.4) does not give sufficient conditions for $d^*$ to be a Bayes decision. For example, local maxima of expected loss will also satisfy this equation. However, if there is a *unique* $d^*$ satisfying equation (7.4) it must be at a (unique) *minimum* and hence be the unique Bayes decision. To see this, note that

$$\bar{S}_b(d) = F(d-b) + 1 - F(d+b)$$

and, from the properties of distribution functions, for fixed values of $b$,

$$F(d-b) \to 0 \quad \text{and} \quad F(d+b) \to \quad \text{as} \quad d \to -\infty$$
$$F(d-b) \to 1 \quad \text{and} \quad F(d+b) \to 1 \quad \text{as} \quad d \to \infty$$

It follows that

$$\bar{S}_b(d) \to 1 \quad \text{as} \quad d \to -\infty \tag{7.6}$$

and

$$\bar{S}_b(d) \to 1 \quad \text{as} \quad d \to \infty \tag{7.7}$$

Since for all $b$

$$0 < \bar{S}_b(d) = 1 - P(d - b \leqslant \theta \leqslant d + b) \leqslant 1$$

it follows that $\bar{S}_b(d)$ has the property that it is decreasing from $d = -\infty$ and increasing as $d \to \infty$ regardless of the form of $F$.

In particular, if $F$ is of a form to have a unique solution of equation (7.4) then the above property of $\bar{S}_b$ ensures that this stationary point attains the unique minimum value of $\bar{S}_b$ and so is the Bayes decision.

If $f(\theta)$ is a linear exponential density (see Theorem 6.1) it can be written in the form

$$f(\theta) = \exp\{\alpha\psi(\theta) + k_2(\alpha)\}$$

Taking logarithms in equation (7.4) we can see that an equivalent equation for Bayes decisions to satisfy is

$$\alpha[\psi(d^* - b) - \psi(d^* + b)] = 0 \tag{7.8}$$

In practice it is often easier to solve for $d^*$ explicitly by using equation (7.8) rather than (7.4).

One of the appealing properties of the step loss function is that whatever a decision-maker's disutility on loss, he must always make the same decision $d^*$.

To see this note that $S_b(\theta - d)$ can only take two values 0 or 1 (say), so $A(S_b(\theta - d))$ can also only take one of two values $a_0$ at 0 and $a_1$ at 1, say, where $a_0 < a_1$, where $A$ is any disutility function on $S_b$. Hence $A$ is effectively identical to the increasing linear transformation $A^*$ where

$$A^*(S_b) = a_0 + (a_1 - a_0)S_b$$

But we know that any increasing linear transformation like $A^*$ gives the same Bayes decision as the one employing an EMV strategy on $S_b$ (see Section 3.4).

Hence your optimal decision or estimate will not depend on your disutility. An estimate will not depend on the preference structure of the decision-maker but only on the beliefs of the decision-maker and the underlying structure of the problem. $S_b$ only shares this property with other loss functions which for any values of $\theta$ and $d$ can only take one of two possible values.

## 7.6 THE GENERALIZED LOCATION OF A RANDOM VARIABLE

*Definition*

Call a real random variable $\theta$ *simple* if its distribution function $F(\theta)$ satisfies the following properties:

(R1) $F(\theta)$ is differentiable with continuous differential $f(\theta)$
(R2) $F(\theta)$ is strictly increasing for all values of $\theta$ such that $F(\theta) > 0$.

The *generalized location* (written $\phi(b)$) of a simple random variable $\theta$ is the set of decisions $d$ indexed by $b$ satisfying:

(G1) $d \notin A$ where $A = \{\theta : F(\theta + \varepsilon) = 0 \text{ for some } \varepsilon > 0\} = (-\infty, a)$ where $a$ is some real number or the empty set.
(G2) $f(\phi(b) - b) = f(\phi(b) + b)$ for some $b > 0$ where $f(\theta)$ is the density of $\theta$.

Most standard distributions are simple. Exceptions include the exponential distribution function, which fails to satisfy R1, and the beta family of distribution functions, which fails to satisfy R2.

The set $\phi(b)$ contains a set of stationary points of $\bar{S}_b(d)$ which contains the Bayes decision with respect to step loss $S_b$. In particular, $\{\phi(b):b>0\}$ contains all decisions which might be Bayes decisions with respect to a step loss function for some value of $b$. To see this, by Theorem 7.3 it is sufficient to show that no stationary point of $\bar{S}_b(d)$ for which $d \in A$ can be a Bayes decision. Suppose $d \in A$ where $A$ is a non-empty set and let $a = \text{Sup } A$. Then

$$\begin{aligned}\bar{S}_b(d) - \bar{S}_b(a) &= F(d-b) + 1 - F(d+b) - (F(a-b) + 1 - F(a+b)) \text{ by (7.5)} \\ &= F(a+b) - F(d+b) \qquad (7.9)\end{aligned}$$

since by definition G1, both $F(d-b)$ and $F(a-b)$ are zero, $b>0$. Furthermore by G2, $F$ is strictly increasing on $(a, \infty)$ and so from equation (7.9),

$$\begin{aligned}&\bar{S}_b(d) - \bar{S}_b(a) \text{ is strictly decreasing in } d, \qquad d > a-b, \text{ for } b>0 \\ &\bar{S}_b(d) = 1, \qquad d < a-b, \text{ by G1}\end{aligned}$$

It follows that if $\theta$ is simple any decision in $A$ is either bettered by $d=a$ or incurs maximum loss, 1. In particular, any stationary point of $\bar{S}_b$ not in $\phi(b)$ are ones for which $S_b(d)=1$. Since $S_a(d) = 1 - F(a+b) < 1$, $b>0$, we also have that for all $d \in A$,

$$\bar{S}_b(d) - \bar{S}_b(a) > 0 \qquad \text{for all } b>0 \qquad (7.10)$$

*Exercise*

Prove that if $f(\theta)$ is unimodal with mode $m$, and $\theta$ is simple, then $\phi(b)$ is a function of $b$ and furthermore

$$\lim_{b \to 0} \phi(b) = m$$

## 7.7 ESTIMATION INTERVALS UNDER ALL 'ESTIMATION' LOSS FUNCTIONS

A loss function $L(\theta, d)$ will be called an *estimation loss function* if it satisfies the following conditions:

(i) $L(\theta, d) = L(\theta - d)$ is a function of $\theta - d$ only;
(ii) $L(\theta, d)$ is symmetric in $\theta - d$;
(iii) $L(\theta, d)$ is bounded above by 1 and below by 0;
(iv) $L(\theta, d)$ is increasing in $|\theta - d|$.

Conditions (i) and (ii) ensure that the loss function depends only on the distance between the estimate and the actual value of $\theta$. Whether the estimate is an overestimate or an underestimate is irrelevant. Condition (iii) ensures that the types of problems mentioned in Section 7.4 are not met. Condition (iv) just says that the more we misestimate by, the more we lose.

The real importance of the step loss function can be seen from the following theorem.

*Theorem 7.4*

Let $L(s) = L(\theta - d)$ be an estimation loss function with

$$\begin{aligned} L(s) &= 0 \quad \text{when} \quad s \leqslant b_1 \\ L(s) &= 1 \quad \text{when} \quad s \geqslant b_2 \end{aligned} \qquad b_1 < b_2$$

Then any Bayes decision with respect to $L$ must lie in the closed interval $[d_1, d_2]$ where

$$\begin{aligned} d_1 &= \inf\{\phi(b): b \in (b_1, b_2)\} \\ d_2 &= \sup\{\phi(b): b \in (b_1, b_2)\} \end{aligned}$$

The proof of this theorem is given in Smith (1980). It is often easy to find $d_1$ and $d_2$ since the calculation of $\phi(b)$ for any specified standard density is usually straightforward. For example, if $\theta$ has a symmetric density with one mode $m$, it is easily shown that $\phi(b) = m$ for all $b > 0$ (see the exercise at the end of Section 7.6). It follows that the EMV Bayes decision under any estimation loss function must be $m$. Here is an example of how to calculate a decision interval when the density of $\theta$ is not symmetric.

*Example 7.3*

Observations $\{X_k : 1 \leqslant k \leqslant n\}$ are independent $X_k$ having a Poisson mass function $P_k(x|\theta)$ where

$$P_k(x|\theta) = \frac{(k\theta)^x}{x} e^{-k\theta} \qquad x = 0, 1, 2, \ldots, \qquad 1 \leqslant k \leqslant n$$

*A priori* your client believes that $\theta$ is distributed gamma $G(\alpha_1, \beta_1)$ with density

$$\pi(\theta) = \frac{\beta_1^{\alpha_1}}{\Gamma(\alpha_1)} \theta^{\alpha_1 - 1} \exp\{-\beta_1 \theta\} \qquad \theta > 0, \text{ where } \alpha_1, \beta_1 > 0$$

If his estimate $d$ of $\theta$ is inaccurate by more than 1 unit he loses all he can. On the other hand, if his estimate $d$ is within $\frac{1}{2}$ unit of $\theta$ he loses nothing. Find an interval in which any sensible estimate $d$ must lie.

**Answer**

$$l(\theta|\mathbf{x}) = \prod_{k=1}^{\theta} \frac{(k\theta)^{x_k}}{(x_k)!} e^{-k\theta} \propto \theta^{n\bar{x}} e^{-r\theta}$$

where

$$r = 1 + 2 + \cdots + n = \tfrac{1}{2}n(n+1) \quad \text{and} \quad \bar{x} = n^{-1} \sum_{i=1}^{n} x_i$$

Thus the posterior density of $\theta$, $p(\theta|\mathbf{x})$, satisfies

$$p(\theta|\mathbf{x}) \propto l(\theta|\mathbf{x})\pi(\theta) \propto \theta^{\alpha-1}e^{-\beta\theta}$$

where

$$\alpha = n\bar{x} + \alpha_1, \qquad \beta = \beta_1 + r \tag{7.11}$$

It follows that $p(\theta|\mathbf{x})$ is a gamma density with parameters $\alpha$, $\beta$ given above. $\phi(b)$ satisfies $f(\phi(b)-b) = f(\phi(b)+b)$ by definition G2 in Section 7.6.

When $f$ is the gamma $G(\alpha, \beta)$ density this becomes, provided $\phi(b) > 0$,

$$\frac{\beta^\alpha}{\Gamma(\alpha)}(\phi(b)-b)^{\alpha-1}\exp\{-\beta[\phi(b)-b)]\}$$

$$= \frac{\beta^\alpha}{\Gamma(\alpha)}(\phi(b)+b)^{\alpha-1}\exp\{-\beta[\phi(b)+b]\}$$

which on taking logs and rearranging gives

$$2\beta b = (\alpha-1)\log\left[\frac{\phi(b)+b}{\phi(b)-b}\right] \tag{7.12}$$

Let $m = (\alpha-1)/\beta$. (It is easily checked that in fact $m$ is the mode of $\theta|\mathbf{x}$.) Equation (7.12) can now be written

$$e^{2m^{-1}b} = 1 + \frac{2b}{\phi(b)-b}$$

i.e.

$$\phi(b) = 2b(e^{2m^{-1}b}-1)^{-1} + b. \tag{7.13}$$

The derivative $\phi'(b)$ of $\phi(b)$ for $b > 0$ is given by

$$\phi'(b) = r(b)[2 + 4m^{-1}br(b)] \qquad \text{where } r(b) = (e^{2m^{-1}b}-1)^{-1} > 0$$

Clearly $\phi'(b)$ is positive when $b > 0$, so $\phi(b)$ is an increasing function of $b$, when $b > 0$. It follows by Theorem 7.4 that, since estimates less than $\frac{1}{2}$ unit out do not matter and estimates more than 1 unit out lose everything, any Bayes decision must lie in the interval

$$[d_1, d_2] = [\phi(\tfrac{1}{2}), \phi(1)]$$

where $\phi(b)$ is given in equation (7.13) and $m = \alpha - 1/\beta$ where $\alpha$ and $\beta$ are given in equation (7.11).

For example, if $m = 1$, $[d_1, d_2] = [1.08, 1.31]$. When $n = \beta_1 = 1$ this interval contains neither the mode or mean of $\theta|\mathbf{x}$.

More examples of the calculation of $\phi(b)$ are given in the exercises at the end of the chapter. The $\phi(b)$ functions associated with some commonly occurring densities are given in Table 7.1.

**Table 7.1** $\phi(b)$ for some standard non-normal densities

| Distribution name | Density proportional to | Mode $m$ | $\phi(b)$ |
|---|---|---|---|
| Gamma | $x^{\alpha-1}e^{-\beta x}$ | $(\alpha-1)\beta^{-1}$ | $b\coth(b/m)$ |
| Log-gamma | $\exp(\alpha x-\beta e^{x})$ | $\log(\alpha/\beta)$ | $m+\log(b/\sinh b)$ |
| Logistic beta | $e^{\alpha x}(1+e^{x})^{-(\alpha+\beta)}$ | $\log(\alpha/\beta)$ | $\log\left[\dfrac{\sinh\{(1+e^{-m})^{-1}b\}}{\sinh\{(1+e^{m})^{-1}b\}}\right]$ |
| Log-normal | $x^{-1}\exp\{-\frac{1}{2}(\gamma+\delta\log x)^2\}$ | $\exp\{-(\gamma\delta^{-1}+\delta^{-2}\}$ | $(m^2+b^2)^{1/2}$ |
| Log $F$ | $e^{(1/2)\alpha x}(\beta+\alpha e^{x})^{-(1/2)(\alpha+\beta)}$ | 0 | $\log\left[\dfrac{(1-\lambda)\sinh(\lambda b)}{\lambda\sinh\{(1-\lambda)b\}}\right], \lambda=\dfrac{\alpha}{\alpha+\beta}$ |

## 7.8 THE UTILITY INVARIANCE OF ESTIMATION INTERVALS

It was pointed out in Section 7.5 that the step loss function had the desirable property that its associated Bayes decision(s) did not vary with the client's disutility function. The estimation intervals of Theorem 7.4 also have the same property. Thus if the decision-maker wishes to make his Bayes decision under any strictly increasing utility function (not just a linear one) on the loss function of the type given in Theorem 7.4 then his Bayes decision must still lie in the region $[d_1, d_2]$ defined in the theorem.

To see this, note that a client will choose his Bayes decision to minimize $E[A(L(d, \theta)]$ where $A$ is a strictly increasing disutility function. But if we let $L^*(d, \theta) = A(L(d, \theta))$, $L^*(d, \theta)$ will still satisfy the conditions of Theorem 7.4. And we know that a Bayes decision under the EMV algorithm, i.e. one that minimizes $[L^*(d, \theta)]$, must lie in $[d_1, d_2]$ by Theorem 7.4. So we can deduce that the preference structure of our client does not influence our choice of interval inside which any decision must lie.

## 7.9 SOME CONCLUDING REMARKS

In this chapter we have studied only the simplest type of decision analysis concerned with more complex statistical models. Unfortunately most interesting decision analytic problems are not so simple. Usually a client will need to make a sequence of decisions in time, and the ramifications of his decisions may be felt well into the future.

Such problems are called *stochastic control problems*. They need somewhat more sophisticated mathematical techniques to identify optimal policies than do non-time dependent problems and in practice usually require numerical analysis on a computer before these policies can be specified precisely. Thus stochastic control analysis is beyond the scope of this introductory text. The interested reader should consult DeGroot (1970), Ross (1983) or Wittle (1983) for an introduction into this fascinating area.

What I hope to have shown in this chapter is that some of the limited class of decision problems considered by statisticians can be embedded within our formal framework. This framework allows for the accommodation of both non-data based information and also the client's needs. Consequent point estimates are therefore more related to the client's problem than are the more traditional methods. Not surprisingly it has been found that Bayesian solutions to common statistical problems are usually different from traditional estimates.

In my opinion, the only interesting statistical problems are those where a client needs to use the results of an analysis to guide the way he *acts*. Such statistical problems are really decision analytic. And the only method of decision analysis which currently has a strong logical foundation, is well understood and is simple enough to provide algorithmic solutions to moderately large decision problems is the Bayesian one. It is therefore difficult not to advocate the use of Bayesian methodology for the solution of statistical decision problems.

## EXERCISES

7.1* Suppose a random variable $X$ with mean $\mu$ has distribution function $F(x)$ with bounded derivatives $f(x)$ and $f'(x)$ respectively. Let random variables $\{Y_n : n > 1\}$ have respective distribution functions $G_n(x)$ given by

$$G_n = (n-1)/nF(x) + n^{-1}H_n(x)$$

Here $\{H_n(x) : n > 1\}$ is any sequence of distribution functions such that

(i) the means $\lambda_n$ of $H_n$ satisfy $\lambda_n = n^2$.
(ii) the first and second derivatives $h_n$ and $h'_n$ of $H_n$ are bounded above by a value $M$ where $M$ does not depend on $n$.

Show that $G_n(x)$, with respective first and second derivatives $g_n(x)$ and $g'_n(x)$, is such that $G_n(x) \to F(x)$, $g_n(x) \to f(x)$ and $g'_n(x) \to f'(x)$ uniformly in $x$ as $n \to \infty$ and yet $|\mu - \mu_n| \to \infty$ as $n \to \infty$. (This proves that arbitrarily 'close' distributions can have widely divergent means. Expectation is thus not a 'stable' decision for the location of a random variable.)

7.2* It is often argued that quadratic loss functions should be used because they are a second-order approximation of any symmetric loss function with second derivative in a neighbourhood of $S$ near 0, where $S = (d - \theta)$. Show that the loss function

$$L(\theta - d) = 1 - \exp\{-\tfrac{1}{2}(\theta - d)^2\}$$

when expanded in its Taylor series is approximately quadratic.

Your posterior density $f(\theta)$ of $\boldsymbol{\theta}|\mathbf{x}$ is given by

$$f(\theta) = \tfrac{1}{2}[f_1(\theta) + f_2(\theta)]$$

where $f_1(\theta)$ and $f_2(\theta)$ are densities of normal random variables with respective means $-8$ and 8 and unit variance. Show that $E(\theta|\mathbf{x})$ is a local *maximum* of the expected loss associated with $L(\theta - d)$ given above, so that by using the approximation above we advocate the use of a 'worst' decision! (For an extension of this argument see Smith, 1979.)

7.3 Let a loss function $L(d - \theta)$ be defined by

$$L(d-\theta) = \begin{cases} 0 & |\theta - d| \leqslant b \\ 1 & (\theta - d) > b \\ B & (\theta - d) < -b \end{cases} \qquad \text{where } B > 1 \text{ and } b > 0.$$

(i) Find the equation giving the minima in (ε) the real line ($R$) of expected loss for a continuous p.d.f. $f(\theta)$ on $\theta$. Give an example of a $t$-distribution arising as a posterior distribution in Bayesian analysis. Suppose.

$$f(\theta) \propto \left(\frac{(r-1)}{2} V + \theta^2\right)^{-r} \qquad V > 0, r > 1$$

Show that the Bayes decision for $\theta$ given the above loss function with $b = 1$ satisfies the equation

$$(d-k)^2 = (k^2 - 1) - \left(\frac{r-1}{2}\right) V \qquad \text{if } d \in R, \text{ where } k = \frac{B^{1/r}+1}{B^{1/r}-1}.$$

Draw the possible shapes of graph of $\bar{L}(d)$ (the expected loss associated with decision $d$) that this generates for different values of $B$, $r$, $V$. Interpret the results.

(ii) Find the Bayes decision when $f(\theta)$ is a normal density and show that in this case using the loss function given above does not give rise to these difficulties.

7.4 Calculate $\phi(b)$ for the following distributions:

(i) $f_1(x)$ unimodal and symmetric about $m$ with $f_1(x) > 0$ for all $x$.

(ii) $f_2(x) \propto \begin{cases} x^{-1} \exp\{-\frac{1}{2}(\gamma + \phi \log x)^2\} & x > 0 \\ 0 & \text{otherwise, } \gamma, \delta > 0. \end{cases}$

(iii) $f_3(x) \propto \exp(-\beta \exp x) \exp \alpha x, \qquad \alpha, \beta > 0.$

Express your answers in terms of the mode $m$ of $f(x)$ and $b$ only. For fixed $m$ draw each $\phi(b)$ for values of $b \leqslant m$.

7.5 (i) Let a distribution function $F(x)$ have density function $f(x)$ that is strictly decreasing and continuous when $x \geqslant a$ and zero when $x < a$. Let the step loss function $S_b$ be defined by

$$S_b(x-d) = \begin{cases} 0 & |x-d| < b, \quad b > 0 \\ 1 & \text{otherwise} \end{cases}$$

By considering the form of the corresponding expected loss function, or otherwise, show that the Bayes decision $d^*$ with respect to $S_b$ and $F$ is unique and satisfies

$$d^* = a + b$$

(ii) The exponential distribution function $F(x)$ is given by

$$F(x) = \begin{cases} 1 - \exp(-\lambda x) & x > 0, \quad \lambda > 0 \\ 0 & \text{otherwise} \end{cases}$$

and a loss function $L(x-d)$ by the equation

$$L(x-d) = S_{b_1}(x-d) + S_{b_2}(x-d)$$

where $S_b$ is defined above and $0 < b_1 < b_2$. Show that, for an arbitrary increasing disutility function $A$ for the decision-maker, one of his Bayes decisions under that disutility is either $b_1$ or $b_2$. Prove, furthermore, that his Bayes decision is uniquely $b_1$ if $(A(2)-A(1))/(A(1)-A(0)) < \alpha$ and is uniquely $b_2$ if $(A(2)-A(1))/(A(1)-A(0)) > \alpha$, where $\alpha = 2\exp(\lambda b_2)\sinh(\lambda b_1)$ and $\lambda$, $b_1$ and $b_2$ are defined above.

7.6 A doctor has to treat a patient who has just been bitten by a snake and now has a quantity $\theta$ of venom in his blood. *A priori* he believes that $\theta$ has a gamma distribution with density $\pi(\theta)$ given by:

$$\pi(\theta) = \begin{cases} 0 & \theta \leqslant 0 \\ \dfrac{\beta^\alpha}{\Gamma(\alpha)}\theta^{\alpha-1}\exp(-\beta\theta) & \theta > 0 \end{cases}$$

where $\alpha = 2$ and $\beta = 3$. To obtain information about $\theta$ he has taken three independent observations $x_1, x_2$ and $x_3$. Each of the random variables $X_1, X_2$ and $X_3$ has an exponential density $f(x|\theta)$ given by:

$$f(x|\theta) = \begin{cases} 0 & x < 0 \\ \theta\exp(-\theta x) & x \geqslant 0 \end{cases}$$

(i) Find his posterior distribution of $\theta$ given $x_1, x_2$ and $x_3$.

(ii) He now has to decide the quantity $d$ of antidote to inject into the patient's bloodstream. Unfortunately, this antidote is itself poisonous so that if too much is administered it is harmful to the patient. It has been discovered that a patient's recovery prospects are represented by some decreasing function $(U|\log d - \log\theta|)$. It is also known that the treatment will have no effect at all if $|\log d - \log\theta| > 1$. Find an interval $[d_1, d_2]$ inside which the quantity of antidote that maximizes the patient's expectation of $U$ must lie whatever the form of $U$. Express $d_1$ and $d_2$ in terms of the sample means of the observations.

(*Hint*: You may assume that the generalized location map for the log-gamma distribution is a decreasing function of its argument $b$ when $0 < b \leqslant 1$).

7.7 Independent measurements $X_1, \ldots, X_n, X_{n+1}$ of the log-concentration of a constituent $A$ of a pharmaceutical product constitute a random sample from a density $f(x|\theta)$ given by

$$f(x|\theta) = \theta\exp\{x - \theta\exp(x)\}, \qquad \theta > 0.$$

Measurements on $(X_1, X_2, \ldots, X_n)$ have been taken, yielding values

$(x_1, x_2, \ldots, x_n)$, in order to forecast the final measurement $X_{n+1}$. The unknown parameter $\theta$ is characteristic of the particular vat from which $X_1, \ldots, X_{n+1}$ are taken. Your beliefs about $\theta$, before any experimentation had taken place, could be summarized by a gamma distribution, with density

$$\pi(\theta) = \frac{\beta^{\alpha}}{\Gamma(\alpha)} \theta^{(\alpha-1)} \exp(-\beta\theta) \qquad \begin{array}{l} \alpha, \beta > 0. \\ \theta > 0. \end{array}$$

Find the posterior density of $\theta$ and the predictive density of $X_{n+1}$ given $x_1, \ldots, x_n$. A new customer will be satisfied with your quality control and so will order from you if and only if your forecast $\hat{x}_{n+1}$ of $X_{n+1}$ satisfies $|\hat{x}_{n+1} - X_{n+1}| < 1$. Find your optimal forecast of $X_{n+1}$, based on $(x_1, x_2, \ldots, x_n)$.

7.8 Prior to taking any observations you feel that the logarithm of the price $\theta$ of a particular product you have to budget for is distributed normally with unit variance i.e. $\log\theta \sim N(m, 1)$. You then observe the prices $Y_i$, $1 \leqslant i \leqslant n$, of $n$ similar products. You believe $Y_1, \ldots, Y_n$ are independent with distribution given by $\log Y_i \sim N(\log\theta, 2)$, $1 \leqslant i < n$.

(i) Find the posterior distribution of the predicted price $\theta$.

(ii) You must make an estimate of $\theta$ and if you underestimate or overestimate by more than 1 unit of price then the firm you work for goes bust. Give an interval in which any sensible decision about $\theta$ must lie.

7.9 (i) The ramp loss function is defined by

$$R_b(\theta - d) = \begin{cases} 1/b|\theta - d| & |\theta - d| \leqslant b \\ 1 & \text{otherwise, } b > 0. \end{cases}$$

Draw $R_b$ for different values of $b$. Show that a Bayes estimate $d^*$ under loss function $R_b$ must satisfy

$$F(d^*) - F(d^* - b) = F(d^* + b) - F(d^*)$$

(ii) Represent this equation pictorially on the density $f$ of $F$ and draw a density which has at least two solutions to the equation above.

(iii) If $F$ has an exponential distribution function use the equation above to express $d^*$ explicitly in terms of $b$ and the rate parameter $\lambda$ of the distribution function $F$.

# Appendix 1

*Lemma A1.1*

Under the notation of Chapter 3, if $r_1, r_2, r^*$ are three rewards such that

$$r_1 \overset{*}{<} r^* \overset{*}{<} r_2 \text{ and you agree to comply with Rules 1–4,}$$

then there exists a unique value of $0 < \alpha < 1$ such that

$$r^* \overset{*}{=} \alpha r_2 + (1-\alpha) r_1 \tag{A1.1}$$

**Proof** First notice that such a value $\alpha$, if it exists, must be unique. For if there exists $\alpha_1 < \alpha_2$ such that

$$\begin{array}{ll} r^* \overset{*}{=} \alpha_1 r_2 + (1-\alpha_1) r_1 & \\ & r_1 \overset{*}{<} r_2 \\ r^* \overset{*}{=} \alpha_2 r_2 + (1-\alpha_2) r_1 & \end{array}$$

then by Rules 1 and 2

$$\gamma r_1 + (1-\gamma) P \overset{*}{=} \gamma r_2 + (1-\gamma) P$$

where

$$\gamma = \alpha_2 - \alpha_1 \quad \text{and} \quad P = \delta r_1 + (1-\delta) r_2, \delta = \alpha_1 (1 + \alpha_1 - \alpha_2)^{-1}$$

But this contradicts Rule 3. So such an $\alpha$, if it exists, must be unique.

Let

$$I_1 = \{\alpha \in [0, 1] : \alpha r_2 + (1-\alpha) r_1 \overset{*}{<} r^*\}$$
$$I_2 = \{\alpha \in [0, 1] : \alpha r_2 + (1-\alpha) r_1 \overset{*}{>} r^*\}$$

Clearly $I_1$ and $I_2$ are both non-empty since $I_1$ contains 0 and $I_2$ contains 1. It follows that since $I_1$ and $I_2$ are bounded,

$$\sigma = \sup I_1 \quad \text{and} \quad \iota = \inf I_2 \text{ both exist}$$

A. We now show that $\sigma \leqslant \iota$, by contradiction. Assume $\sigma > \iota$, then there exist $s \in I_1$ and $i \in I_2$ such that $s > i$. So there exists $s > i$ such that

$$s r_2 + (1-s) r_1 \overset{*}{<} r^* \overset{*}{<} i r_2 + (1-i) r_1 \tag{A1.2}$$

Let $P = \beta r_2 + (1-\beta) r_1$ where $\beta = i(1-s+i)^{-1}$

Then by Rule 1

$$s r_2 + (1-s) r_1 \overset{*}{=} \gamma r_2 + (1-\gamma) P$$

and

$$ir_2 + (1-i)r_1 \overset{*}{=} \gamma r_1 + (1-\gamma)P \qquad \text{where } \gamma = s - i > 0.$$

So by Rule 2 and equation (A1.2)

$$\gamma r_2 + (1-\gamma)P \overset{*}{<} \gamma r_1 + (1-\gamma)P \qquad \text{where } r_1 \overset{*}{<} r_2$$

This clearly contradicts Rule 3. We can therefore conclude that $\sigma \leqslant \iota$. In fact $\sigma = \iota$, for if $\sigma < \iota$, by Rule 2, all $\sigma \in (\sigma, \iota)$ must satisfy equation (A1.1) which contradicts the uniqueness of such an $\alpha$,

So if the required $\alpha$ exists then

$$\alpha = \sigma = \iota.$$

B. We must therefore finally prove that $\sigma \notin I_1$ and $\iota \notin I_2$. For then by Rule 2, $\alpha$ must satisfy equation (A1.1). If $\sigma \in I_1$, let

$$P_1 = r_2$$
$$P_2 = \sigma r_2 + (1-\sigma)r_1$$
$$P = r^*$$

Then by Rule 4 there exists a $\beta \in (0, 1)$ such that

$$r^* \overset{*}{>} \beta P_2 + (1-\beta)r_2 \qquad \text{which by Rule 1}$$
$$\overset{*}{=} \delta r_2 + (1-\delta)r_1$$

where $\delta = (1-\beta) + \beta\sigma > \sigma$ since $\beta < 1$ and $\sigma \neq 1$ because by hypothesis $\sigma \in I_1$ and we know $1 \in I_2$. On the other hand, if $\iota \in I_2$ let

$$P_1 = \iota r_2 + (1-\iota)r_1$$
$$P_2 = r_1$$
$$P = r^*$$

Then, by Rule 4, there exists an $\alpha \in (0, 1)$ such that

$$r^* \overset{*}{<} \alpha r_1 + (1-\alpha)P_1 \qquad \text{which by Rule 1}$$
$$\overset{*}{=} \gamma r_2 + (1-\gamma)r_1$$

where $\gamma = (1-\alpha)\iota < \iota$ since $\alpha > 0$ and $\iota \neq 0$ because by hypothesis $\iota \in I_2$ and we know $0 \in I_1$.

Hence $\sigma \notin I_1$ and $\iota \notin I_2$. So there exists $\alpha = \sigma$ satisfying equation (A1.1).

*Lemma A1.2*

Let $P$ have a set of rewards $r_1, r_2, \ldots, r_n$ on which it assigns non-zero probability $p(r_i)$ to $r_i$, $1 \leqslant i \leqslant n$.
Then

$$P \overset{*}{=} Q \quad \text{where} \quad Q = \beta t + (1-\beta)s$$

where $s$ and $t$ are respectively a least preferable and a most preferable possible reward and

$$\beta = \sum_{i=1}^{n} p(r_i)\alpha(r_i)$$

where $\alpha(r_i)$ is the unique (by Lemma A1.1) probability satisfying

$$r_i \overset{*}{=} \alpha(r_i)t + (1 - \alpha(r_i))s = Q_i \qquad \text{(say)}$$

**Proof** Let $P_i$ be the probability distribution on $r_i, r_{i+1}, \ldots, r_n$ which assigns

$$p_i(r_k) = (1 - \sum_{j=1}^{i-1} p(r_j))^{-1} p(r_k) \qquad i \leqslant k \leqslant n, 2 \leqslant i \leqslant n$$

$$\begin{aligned}
P &= p(r_1)r_1 + (1 - p(r_1))P_2 \\
&\overset{*}{=} p(r_1)Q_1 + (1 - p(r_1))P_2 \qquad \text{by Rule 3} \\
&= p(r_1)Q_1 + p_2(r_2)r_2 + (1 - p(r_1) - p(r_2))P_3 \\
&= p(r_2)r_2 + (1 - p(r_2))R_1 \qquad \text{where } R_1 = \gamma Q_1 + (1-\gamma)P_3 \\
&\qquad\qquad\qquad\qquad\qquad \text{where } \gamma = (1 - p(r_2))^{-1}(r_1) \\
&\overset{*}{=} p(r_2)Q_2 + (1 - p(r_2))R_1 \qquad \text{by Rule 3} \\
&= p(r_1)Q_1 + p(r_2)Q_2 + (1 - p(r_1) - p(r_2))P_3 \\
&\overset{*}{=} \sum_{i=1}^{n} p(r_i)Q \\
&= \beta t + (1-\beta)s \qquad \text{by Rule 1, where } \beta \text{ is defined above.}
\end{aligned}$$

*Lemma A1.3*

Assume Rules 1–4 hold. Suppose there exists a least favourable reward $s$ such that for all possible rewards $r$ either $s \overset{*}{<} r$ or $s \overset{*}{=} r$. In addition, assume that there exists a most favourable reward $t$ such that for all other possible rewards $r$, either $t \overset{*}{>} r$ or $t \overset{*}{=} r$. Then the only possible utility functions expressing your preferences must be of the form

$$U(r) = a\beta + b \qquad a > 0 \tag{A1.3}$$

where $\beta \in [0, 1]$ is the unique (by Lemma A1.1) probability such that

$$r \overset{*}{=} \beta t + (1-\beta)s$$

Furthermore, under such a definition, for all $r, r_1, r_2$ such that $r_1 \overset{*}{<} r \overset{*}{<} r_2$, with $r \overset{*}{=} \alpha r_2 + (1-\alpha)r_1$,

$$U(r) = \alpha U(r_2) + (1-\alpha)U(r_1) \tag{A1.4}$$

**Proof** If $U(r)$ is a utility function, in particular it must satisfy equation (A1.4). So in particular, putting $r_1 = s$, $r_2 = t$ we must have that

$$U(r) = \beta U(t) + (1-\beta)U(s)$$
$$= a\beta + b \qquad \text{where } a = U(t) - U(s)$$
$$b = U(s) \qquad \text{(A1.5)}$$

The function $U(r)$ must also have the property that

$$r_1 \overset{*}{<} r_2 \Leftrightarrow U(r_1) < U(r_2)$$

This implies that in equation (A1.3), $a = U(t) - U(s) > 0$. So if $U$ is a utility function it must have the form (A1.3).

We now show that any function $U$ satisfying (A1.3) acts to satisfy (A1.4). Under our definition

$$U(r_i) = a\alpha_i + b \qquad i = 1, 2$$

where $\alpha_i$ is any probability satisfying $r_i \overset{*}{=} \alpha_i t + (1-\alpha_i)s$. If $r_1 = s$ then uniquely $\alpha_1 = 0$ and if $r_2 = t$ uniquely, $\alpha_2 = 1$ by the definition of $a$ and $b$ in (A1.5) and by Rule 3. On the other hand, if $0 < \alpha_i < 1$, then it is unique by Lemma A1.1.

Also $U(r) = a\beta + b$, where $r \overset{*}{=} \beta t + (1-\beta)s$.

Now suppose that $\alpha \in (0, 1)$ is such that

$$r \overset{*}{=} \alpha r_2 + (1-\alpha) r_1$$

By Lemma A1.2 we can now deduce that

$$\beta = \alpha\alpha_2 + (1-\alpha)\alpha_1$$

Hence

$$U(r) = a\beta + b = a(\alpha\alpha_2 + (1-\alpha)\alpha_1) + b$$
$$= \alpha(a\alpha_2 + b) + (1-\alpha)(a\alpha_1 + b)$$

So

$$U(r) = \alpha U(r_2) + (1-\alpha)U(r_1) \qquad \text{(A1.6)}$$

Finally, from (A1.6), for any $r_1 \overset{*}{<} r \overset{*}{<} r_2$

$$U(r) = [U(r_2) - U(r_1)]\alpha + U(r_1)$$

so that $0 < a = U(r_2) - U(r_1)$. Thus, in particular, if $r_1 \overset{*}{<} r_2$

$$U(r_1) < U(r_2)$$

our second requirement of a utility function. This completes the proof.

*Theorem A1.1*

There exists a utility function $U$ such that for any two distributions of rewards $P_1$ and $P_2$ which assign zero probability to all but a finite number of rewards

$$P_1 \overset{*}{<} P_2 \Leftrightarrow \bar{U}(P_1) < \bar{U}(P_2)$$

where $\bar{U}(P)$ denotes the expectation of $U$ with respect to $P$.

**Proof** Let $r_1, r_2, \ldots, r_n$ denote the set of rewards that are assigned a non-zero probability by either $P_1$ or $P_2$, and let $p_j(r_i)$ denote the probability assigned to $r_i$ by $P_j$, $1 \leqslant i \leqslant n$, $j = 1, 2$.
Lemma A1.2 proved that

$$P_j \stackrel{*}{=} Q_j \quad \text{where } Q_j = \beta_j t + (1 - \beta_j)s \tag{A1.7}$$

where $s$ and $t$ are respectively the minimum and maximum possible reward and

$$\beta_j = \sum_{i=1}^{n} p_j(r_i)\alpha(r_i)$$

where $\alpha(r_i)$ is the unique (by Lemma A1.1) probability satisfying

$$r_i \stackrel{*}{=} \alpha(r_i)t + (1 - \alpha(r_i))s$$

Now use the definition of $U$ given in Section 3.3.

$$\beta_j = \sum_{i=1}^{n} p_j(r_i)U(r_i) = \bar{U}(P_j) \qquad \text{by definition of expectation and } U(r_i) \tag{A1.8}$$

By equation (A1.7) and Rules 1 and 2,

$$Q_1 \stackrel{*}{<} Q_2$$

which by Rule 3 implies

$$\begin{aligned} &\beta_1 < \beta_2 \\ &Q_2 = \delta t + (1 - \delta)P \qquad t \stackrel{*}{>} s \qquad && \delta = \beta_2 - \beta_1 \\ &Q_1 = \delta s + (1 - \delta)P && P = \gamma t + (1 - \gamma)s \\ & && \text{where } \gamma = (1 - \beta_2 + \beta_1)^{-1}\beta_1 \end{aligned}$$

which by (A1.8) implies

$$\bar{U}(P_1) < \bar{U}(P_2).$$

Since this holds for any two arbitrary distributions $P_1$ and $P_2$ our theorem is proved.

# Appendix 2

*Theorem A2.1*

Any influence diagram satisfying conditions (a) and (b) of the definition of extensive form (Section 5.2.2) has at least one node whose only direct successor is the value node $n(v)$.

**Proof** Assume to the contrary that every direct predecessor of $n(v)$ has at least one other direct successor. Form a new graph $I'$ by deleting $n(v)$ from $I$ together with all arcs into it.

By the above assumption and (b), every node in $I'$ has at least one direct successor. Starting with any node in $I'$ a path can be traced along the directed arcs in $I'$. Since each node has a direct successor, such a path need not terminate until we return to a node on this path. So $I'$ contains at least one cycle. But $I'$ is a subgraph of $I$ and so $I$ contains at least one cycle, in contradiction to condition (b). So we have proved that at least one node in $I$ has as its only direct successor the value node $n(v)$.

*Theorem A2.2*

Let $I$ be an influence diagram in extensive form for which no chance node has the value node $n(v)$ as its only direct successor. Then every chance node in $I$ is a direct predecessor of $D_m$, where $D_m$ is defined in condition (d) of Section 5.2.2.

**Proof** Assume to the contrary that there exists a chance node $n(X_1)$ that is not a direct predecessor of $D_m$. If $n(X_1)$ is the direct predecessor of any decision node, then by condition (d) there would be a path from $n(X_1)$ to $D_m$, in contradiction to condition (c). By condition (b) and the condition of the theorem, $n(X_1)$ must have a direct successor which is not the value node $n(v)$. So there must be another chance node $n(X_2)$ that directly succeeds $n(X_1)$.

Now clearly $n(X_2)$ is not a direct predecessor of $D_m$ for then condition (c) is violated. By the argument of the previous paragraph $n(X_2)$ cannot have a decision node as a direct successor. So by condition (b) and the condition of the theorem again, $n(X_2)$ must have a direct predecessor $n(X_3)$.

Exactly the same argument gives us that $n(X_3)$ must have a direct successor chance node $n(X_4)$, $n(X_4)$ a direct successor chance node $n(X_5)$, and so on. Since $I$ only contains a finite number of chance nodes it follows that $I$ must contain a cycle among its chance nodes. But this contradicts condition (a) and we have our required contradiction.

*Theorem A2.3*

If an influence diagram $I$ is in extensive form, then the influence diagram $I_1$ constructed in Section 5.2.2 must also be in extensive form.

**Proof** If condition (b) holds for $I$ it must hold for $I_1$ too, because under our construction the number of direct successor nodes other than nodes $n(1)$ and $n(v)$ is the same for each graph. And clearly $n_1(v)$ has no direct successor by our construction. So diagram $I_1$ satisfies condition (b) of our definition.

If $I_1$ had a directed cycle, this cycle could not go through $n_1(v)$ because $n_1(v)$ has no direct successors. So this cycle would have to go through nodes other than $n_1(v)$. But the arcs between these nodes are the same as they were in $I$ and $I$ has no directed cycle by definition. So condition (a) holds in $I$.

Condition (c) holds for $I_1$ if it holds in $I$ because our construction adds no new arcs between chance nodes originally in $I$ nor deletes arcs from chance nodes to decision nodes. Condition (d) continues to hold in $I_1$ because no arcs between decision nodes remaining in $I_1$ are deleted by our re-representation.

# Appendix 3

*Theorem A3.1*

Under absolute loss given in equation (7.1) a decision $d$ is a Bayes decision if and only if it is an $a_1(a_1+a_2)^{-1}$ percentile of $\theta$, provided that $E|\theta| = M$ where $M$ is finite.

**Proof** Since $A(d,\theta)$ is a function of $d-\theta$ only, without loss write $A(d,\theta) = A(d-\theta)$.

We first show that if $E|\theta|$ exists then the expected loss $\bar{A}(d)$ associated with decision $d$ and parameter $\theta$ exists. To see this note that

$$\begin{aligned} 0 < A(\theta, d) &\leqslant a^*|\theta - d| && \text{where } a^* = \max\{a_1, a_2\} \\ &\leqslant a^*(|\theta| + |d|) && \text{by the triangle inequality} \end{aligned}$$

So $$0 < \bar{A}(d) \leqslant a^*(M + |d|) \qquad \text{where this bound is finite.}$$

Hence the expected loss exists for all values of $d$.
Let $d^*$ be an $a_1(a_1+a_2)^{-1}$ percentile.

*Case I* $d > d^*$

$$A(d-\theta) - A(d^* - \theta) = \begin{cases} a_2(d-d^*) & \theta \leqslant d^* \\ r(\theta) - a_1(d-d^*) & d^* < \theta < d, \\ -a_1(d-d^*) & d < \theta \end{cases}$$

$$\text{where } r(\theta) = (a_1+a_2)(d-\theta) > 0$$

Let

$$r'(\theta) = \begin{cases} r(\theta) & d^* < \theta < d \\ 0 & \text{otherwise} \end{cases}$$

Then

$$\begin{aligned} \bar{A}(d) - \bar{A}(d^*) &= E[A(d-\theta) - A(d^* - \theta)] \\ &= a_2(d-d^*)P(\theta \leqslant d^*) - a_1(d-d^*)[1 - P(\theta \leqslant d^*)] + E(r'(\theta)) \end{aligned}$$

which, since $r'(\theta) \geqslant 0$

$$\geqslant (d - d^*)[(a_1 + a_2)P(\theta \leqslant d^*) - a_1]$$

$$\geqslant (d - d^*)(a_1 + a_2)\left[\frac{a_1}{a_1 + a_2} - a_1\right]$$

since $d^*$ is an $a_1(a_1 + a_2)^{-1}$ percentile of $\theta = 0$

$$\bar{A}(d) \geqslant \bar{A}(d^*) \text{ with equality } \quad \text{only if } E(r'(\theta)) = 0 \qquad \text{(A3.1.)}$$

*Case II* $d < d^*$

$$A(d - \theta) - A(d^* - \theta) = \begin{cases} -a_2(d - d^*) & \theta \leqslant d \\ -s(\theta) + a_2(d_1 - d^*) & d < \theta < d^*, \\ a_1(d - d^*) & d^* \leqslant \theta \end{cases}$$

where $$s(\theta) = (a_1 + a_2)(d^* - \theta) > 0$$

Let

$$s'(\theta) = \begin{cases} s(\theta) & d < \theta < d^* \\ 0 & \text{otherwise} \end{cases}$$

Then

$$\bar{A}(d) - \bar{A}(d^*) = E[A(d - \theta) - A(d^* - \theta)]$$

$$= -a_2(d - d^*)[1 - P(\theta \geqslant d^*)] + a_1(d - d^*)P(\theta \geqslant d^*) + E(s'(\theta))$$

which since $s'(\theta) \geqslant 0$

$$\geqslant (d - d^*)[(a_1 + a_2)P(\theta \geqslant d^*) - a_2]$$

$\geqslant 0$ since $d^*$ is an $a_1(a_1 + a_2)^{-1}$ percentile of $\theta$

$$\bar{A}(d) \geqslant \bar{A}(d^*) \text{ with equality } \quad \text{only if } E(s'(\theta)) = 0 \qquad \text{(A3.2)}$$

This proves that any $a_1(a_1 + a_2)^{-1}$ percentile $d^*$ is a Bayes decision.

Finally note that if $d < d^*$, since $r(\theta) > 0$, $d^* < \theta < d$

$$A(d) = A(d^*) \Rightarrow E(r'(\theta)) = 0 \quad \text{by equation (A3.1)}$$

$$\Rightarrow P(d^* < \theta < d) = 0$$

$\Rightarrow d$ is another $a_1(a_1 + a_2)^{-1}$ percentile of $\theta$.

Similarly, if $d > d^*$,

$$A(d) = A(d^*) \Rightarrow E(s'(\theta)) = 0 \quad \text{by equation (A3.2)}$$

$$\Rightarrow P(d < \theta < d^*) = 0$$

$\Rightarrow d$ is another $a_1(a_1 + a_2)^{-1}$ percentile of $\theta$.

This proves that the *only* Bayes decisions are $a_1(a_1 + a_2)^{-1}$ percentiles of $\theta$.

# References


Aitchison, J. and Dunsmore, I.R. (1975) *Statistical Prediction Analysis*, Cambridge University Press, Cambridge.

Arrow, K.J. (1951) *Social Choice and Individual Values*, Wiley, New York.

Bacharach, M. (1975) Group decisions in the face of differences of opinion, *Management Science*, **22**, 182–91.

Bell, D.E., Keeney, R.L. and Raiffa, H. (1977) *Conflicting Objections in Decision*, Wiley, Chichester.

Craven, B.D. (1981) *Functions of Several Variables*, Chapman and Hall, New York.

Darroch, J.N., Lauritzen, S.L. and Speed, T.P. (1980) Markov fields and loglinear models for contingency tables, *Annals of Statistics*, **8**, 522–39.

Davidson, D., Suppes, P. and Siegel, S. (1957) *Decision Making: An Experimental Approach*, Stanford University Press, Stanford, California.

Dawid, A.P. (1982) The well-calibrated Bayesian, *Journal of the American Statistical Association*, **77**, 605–10.

Dawid, A.P. (1983) *Probability Forecasts*, University College London Research Report.

Dawid, A.P. (1984) *Calibration-Based Empirical Probability*, University College Statistical Science Research Report No. 36.

De Finetti, B. (1974) *Theory of Probability*, Vols 1 and 2, J. Wiley, New York.

DeGroot, M.H. (1970) *Optimal Statistical Decisions*, McGraw-Hill, New York.

DeGroot, M.H. and Fienberg, S.E. (1983) The comparison and evaluation of forecasters, *The Statistician*, **32**, 12–22.

Fine, T.L. (1973) *Theories of Probability*, Academic Press, New York.

French, S. (1985) Group consensus probability distributions: a critical survey, *Proceedings of the Second Valencia International Meeting* (Sept. 1983) (Ed. Bernado, DeGroot, Lindley and Smith), pp. 183–202.

French, S. (1986) *Decision Theory; an Introduction to the Mathematics of Rationality*, Ellis Horwood, Chichester.

Fullerton, G.H. (1971) *Mathematical Analysis*, University Mathematical Texts.

Genest, G. and Zidek, J.V. (1986) Combining probability distributions: a critique and an annotated bibliography, *Statistical Science*, **1**(1), 114–48.

Gibbard, A. (1973) Manipulation of voting schemes: a general result, *Econometrics*, **41**, 587–601.

Goldstein, M. (1981) Revising previsions: a geometric approach (with discussion), *J.R. Statist. Soc.*, B, **43**(2) 105–30.

Good, I.J. (1950) *Probability and the Weighting of Evidence*, Griffin, London.

Good, I.J. (1983) Weight of evidence: a brief survey, Invited paper: 2nd International Meeting on Bayesian Statistics, Valencia.

Harrison, P.J. and Stevens, C.F. (1976) Bayesian Forecasting (with discussion) *J.R. Statist. Soc.*, B, **38**, 205–47.

Howard, R.A. and Matheson, J.E. (1981) Influence diagrams, in *The Principles and Applications of Decision Analysis*, Vol. II (ed. Howard, R.A. and Matheson, J.E., 1984), Strategic Decisions Group, Menlo Park, Calif.

Humphreys, P. (1977) Application of multi-attribute utility theory, in *Decision Making and Change in Human Affairs*, Reidel, pp. 165–209.

Kadane, J.B. and Chuang, D.T. (1978) Stable decision problems, *Ann. Statist.*, **6**, 1095–111.

Keeney, R.L. and Raiffa, H. (1976) *Decisions with Multiple Objectives: Preferences and Value Trade-offs*, Wiley, New York.

Kim, J.H. and Pearl, J. (1983) A computational model for causal and diagnostic reasoning in inference systems, *Proceedings 8th International Joint Conference on Artificial Intelligence*, Karlsruhe, West Germany, pp. 190–3.

Lichtenstein, S. and Fischhoff, B. (1976) Do those who know more also know more about how much they know?, *Oregon Research Institute Research Bulletin*, **16**(1).

Lichtenstein, S. Fischhoff, B. and Phillips, L.D. (1977) Calibration of probabilities: the state of the art, in *Decision Making and Change in Human Affairs*, Reidel, pp. 275–325.

Litchenstein, S., Fischoff, B. and Phillips, L.C. (1982) Calibration of probabilities: the state of the art in 1980, in Kalmeman, D., Sloviz, P. and Tversky, A. (eds), *Judgement under Uncertainty: Heuristics and Biases*, Cambridge University Press, New York.

Lindley, D.V. (1980) *Introduction to Probability and Statistics from a Bayesian Viewpoint*, Part 2, *Inference*, Cambridge University Press.

Lindley, D.V. (1985a) Reconciliation of discrete probability distributions, *Proceedings of the Second Valencia International Meeting* (Sept. 1983) (Ed. Bernardo, DeGroot, Lindley and Smith), pp. 375–90.

Lindley, D.V. (1985b) *Making Decisions*, Wiley.

Lindley, D.V. and Smith, A.M.F. (1972) Bayes estimates for the linear model (with discussion), *J.R. Statist. Soc.*, B, **34**, 1–41.

Lindley, D.V., Tversky, A. and Brown, R.V. (1979) On the reconciliation of probability assessments, *J.R. Statist. Soc.*, A, **112**, 146–80.

Murphy, A.H. (1973) A new vector partition of the probability score, *Journal of Applied Metrology*, **12**, 595–600.

Nash, J.F. (1950) The bargaining problem, *Econometrica*, **18**, 155–62.

Naylor, J.C. and Shaw, J.E.H. (1985) *Bayes Four User Guide*, Nottingham Statistics Group, Department of Mathematics, University of Nottingham, Nottingham, UK.

Naylor, J.C. and Shaw, J.E.H. (1986) *Bayes Four Implementation Guide*, Nottingham Statistics Group, Department of Mathematics, University of Nottingham, Nottingham, UK.

Pearl, J. (1982) Reverend Bayes on inference engines: a distributed hierarchical approach, *Proceedings Second Annual Conference on Artificial Intelligence*, Pittsburgh, Pennsylvania, pp. 133–6.

Phillips, L.D. (1979) Introduction to decision analysis, *Tutorial Paper 79–1*, Decision Analysis Unit, London School of Economics.

Phillips, L.D. (1982) Generation theory, in *Research Marketing*, Supplement 1: *Choice Models for Bayes Behaviour* (Ed. L. McAlister), JAI Press, Greenwich, Conn.

Phillips, L.D. and Wright, C.N. (1977) Cultural differences in viewing uncertainties and assessing probabilities, in *Decision Making and Change in Human Affairs*, Reidel, pp. 507–21.

Press, S.J. (1972) *Applied Multivariate Analysis*, Holt, Rinehart and Winston, Chicago.

Raiffa, H. (1968) *DecisionAnalysis*, Addison Wesley, Reading, Mass.

Raiffa, H. and Schlaifer, R. (1961) *Applied Statistical Decision Theory*, MIT Press, Cambridge, Mass.

Ross, M.S. (1983) *Introduction to Stochastic Dynamic Programming*, Academic Press.

Savage, L.J. (1954) *The Foundations of Statistics*, Wiley, New York.

Savage, L.J. (1971) Elicitation of personal probabilities and expectations, *J.A.S.A.*, **66**, 783–801.

Schmeidler, D. (1984) Subjective probability and expected utility without additivity, *Research Report*, The Foerder Institute for Economic Research, Tel Aviv University, Israel.

Shachter, R.D. (1984) Evaluating influence diagrams, Department of Engineering – Economic Systems, *Stanford University Research Report CA 94305*. To appear in *Operations Research.*

Shachter, R.D. (1986) Probabilistic inference and influence diagrams. To appear in *Operations Research.*

Shafer, G. (1982) Belief functions and parametric models (with discussion), *J.R. Statist. Soc.*, B, **44**, 322–52.

Smith, J.Q. (1978) Problems in Bayesian statistics relation to discontinuous phenomena, catastrophe theory and forecasting, Ph.D thesis, University of Warwick.

Smith, J.Q. (1980) Bayes estimates under bounded loss, *Biometrika*, **67**(3), 629–38.

Smith, J.Q. (1986) Diagrams of influence in statistical models, Department of Statistics, University of Warwick, *Research Report 99.*

Smith, J.Q. (1987a) Influence diagrams for Bayesian decision analysis, Department of Statistics, University of Warwick, *Research Report 100.*

Smith, J.Q. (1987b) Manipulating influence diagrams, Department of Statistics, University of Warwick *Research Report.*

Spiegelhalter, D.J. (1986a) A statistical view of uncertainty in expert systems, in *Artificial Intelligence and Statistics* (Ed. W. Gale), Addison-Wesley.

Spiegelhalter, D.J. (1986b) Probabilistic reasoning in predictive expert systems. To appear in *Uncertainty in Artificial Intelligence* (Eds L.N. Kaval and J. Lemmer), North-Holland, Amsterdam.

Spiegelhalter, D.J. and Knill-Jones, R.P. (1984) Statistical and knowledge based approaches to clinical decision-support systems, with an application to gastroenterology (with discussion), *J.R.S.S.* A, **147**, 35–78.

United States Weather Bureau (1969) Report on weather bureau forecasts performance 1967–8 and comparison with previous years. *Technical Memorandum WBTMFCST* **11**, Office of Metereological Operations Weather Analysis and Prediction Division.

Walley, P. (1982) The elicitation and aggregation of beliefs, *University of Warwick Research Report.*

West, M. (1984) Bayesian aggregation, *J.R. Statist. Soc.*, A, **147**, 600–607.

Winkler, R.C. and Murphy, A.H. (1968) Evaluation of subjective precipitation forecasts, *Proceedings of First National Conference on Statistical Meterology*, Hartford, Conn., Amer. Meterological Society.

Wittle, P. (1983) *Optimisation over Time*, Vols 1 and 2, Wiley.

Wolfenson, M. and Fine, T.L. (1982) Bayes-like decision making with upper and lower probabilities, *Journal of the American Statistical Association*, **77**, 82–7.

Zidek, J.V. (1983) Multi-Bayesianity-consensus of opinion, Unpublished manuscript, University of London.

# Index